Astronomical Telescopes and Observatories for Amateurs

BOOKS BY PATRICK MOORE

A Handbook of Practical Amateur Astronomy
The Planets
A Survey of the Moon
Naked-Eye Astronomy
The New Look of the Universe
The Amateur Astronomer's Glossary
Amateur Astronomy
The Sun
Suns, Myths, and Men
*How to Make and Use a Telescope, by Patrick Moore and H. P.
    Wilkins*
Life on Mars, by Patrick Moore and Francis Jackson
Craters of the Moon, by Patrick Moore and Peter Cattermole
Can You Speak Venusian
The Yearbook of Astronomy (annual)
Astronomical Telescopes and Observatories for Amateurs

Astronomical Telescopes and Observatories for Amateurs

Edited by **PATRICK MOORE**

CONTRIBUTORS

Patrick Moore
Iain Nicolson
G. A. Hole
T. W. Rackham
Terence Moseley

H. K. Robin
Gilbert E. Satterthwaite
H. E. Dall
J. C. D. Marsh
Colin A. Ronan

H. R. Hatfield
L. Wilson
Henry Brinton
F. R. Spry
J. Hedley Robinson

W · W · NORTON & COMPANY · INC · New York

ISBN   0   393   06395   X

ALL RIGHTS RESERVED
Published simultaneously in Canada
by George J. McLeod Limited, Toronto

PRINTED IN THE UNITED STATES OF AMERICA

1   2   3   4   5   6   7   8   9   0

Contents

Illustrations

7

PLATES

FIGURES

1 : PATRICK MOORE

Introduction

THIS IS ESSENTIALLY A PRACTICAL BOOK, WRITTEN FOR PRACTICAL amateur astronomers by astronomers who are expert in their own fields of study. An attempt has been made to cover a very wide range of subjects, and inevitably the treatment cannot be full; but the essentials are here, and we have also discussed amateur observatories in some detail.

As Editor, I may perhaps be allowed to begin by saying something about the choice of a telescope, and the rôle of the amateur in modern astronomical science. Of course these views are bound to be personal ones, but I hope that they are also logical.

Anyone who aspires to become a professional astronomer must have a recognised degree—if not in pure astronomy, then in physics or perhaps mathematics. This cannot be too strongly emphasised, and these days there is no short cut. But if he is not academically minded, he can still make a serious hobby out of astronomy and make himself scientifically useful, as many modern amateurs have done. The names of Leslie Peltier in America, the young comet discoverers in Japan, W. E. Bennett in South Africa and George Alcock and W. M. Baxter in Britain come instinctively to mind; but there are many others too.

Neither is it true to say that the amateur must be equipped with complicated and very expensive apparatus. Of course a large telescope is very desirable—at least for most fields of study; and if bought new it is a costly item. On the other hand, the expense is non-recurring, and in any case the really skilful

11

'handyman' can make a powerful telescope for himself, even to the extent of grinding the mirror. If so, the total outlay will be much less than that for a colour television set.

I admit, without reservation, that I am not gifted in this respect. I wish I were. It has been said, probably with truth, that I am the clumsiest person in three continents. That being so, I had to have my telescopes made for me; and the problems which I faced when setting out—which was, admittedly, in the pre-war days when prices were far lower than they are now—are still those to be faced today.

In making a start in astronomy on an amateur basis (or, for that matter, with the eventual aim of taking a degree), no equipment whatsoever is needed apart from a pair of eyes. I believe, firmly, that after a certain amount of elementary reading has been done, the best course is to go outside on a starry night, equipped with an outline chart, and begin trying some star recognition. Actually the constellations are not nearly so difficult to identify as might be thought at first glance, and a few evenings' practice will work wonders. After only a few hours outdoors the main groups will be as familiar as the roads in one's home town or village; and the stars become so much more fascinating when one knows which is which.

The next step, for most people, is to obtain some sort of optical equipment; and it is here that the trouble sometimes starts. The beginner consults various catalogues, including some which are excellently produced and which look so very much better than they really are. What sized telescope is needed? A small one, costing a reasonable sum, or else a larger one, which is much more expensive? Photographs in catalogues can often disprove the dictum that the camera cannot deceive! The novice cannot be expected to tell what is worth buying and what should be shunned.

Personally, I believe that the best first step is not to buy a telescope at all, but to invest in a pair of binoculars. These have most of the advantages of very small telescopes, with few of the disadvantages. They are easy to hold, and give a wide field; though the magnification is not high, it is high enough to give superb views of star-fields and clusters, plus features such as the phases of Venus and the mountains and craters of the Moon. A small telescope will give a larger magnification, but a much smaller field, and will be much less easy to handle. (Also, binoculars can be used for more mundane pastimes such as bird-watching.)

For an all-purpose pair of binoculars, I would recommend 7 × 50 (that is to say, a magnification of 7 and object-glasses each 50 millimetres across). I find these very useful, and they are invaluable for some kinds of observing, such as studying bright variable stars. I also have a couple of more powerful pairs: 12 × 60 and 20 × 70, but the latter at least can hardly be used without a mounting, and they cost more. A small outlay will certainly provide a good pair of 7 × 50 binoculars, and second-hand pairs can often be picked up very cheaply.

When we come to telescopes, I admit that I would always be reluctant to pay much money for any refractor below 3in aperture or any reflector below 6in. I would prefer to make do with binoculars, and save up for a telescope which will be powerful enough for serious observation as well as casual star-gazing. This is not to say that smaller instruments, such as 4in reflectors or the 2½in Japanese refractors now on common sale, are useless. They are not; far from it—and they will give immense enjoyment. But their light-grasp is bound to be limited by their small aperture, however good their optical quality. I will not labour the point, and I know that many people will disagree with me; but I can only state my own opinions.

The choice of instrument depends upon one's personal circumstances. The city-dweller is at an immediate disadvantage, because the night sky will be so light that he will be able to see very little; this means either keeping to binoculars, making do with a very small portable telescope, or else abandoning the night-time altogether and concentrating on the Sun—as the late

W. M. Baxter did from his observatory at Acton, London. Poking a telescope out of a window is emphatically not to be recommended, though in some cases there is no alternative.

Even people who live well away from towns find that there are hazards in the form of artificial lights, high buildings or trees. According to the principle of Spode's Law, 'if things *can* be awkward, they *are*', these hazards always lie in the most inconvenient possible direction. Again it may be essential to have a telescope which can be shifted about, and this means that the aperture cannot be great; a 4in refractor is barely portable, and an 8½in Newtonian reflector even less so. The solar, lunar or planetary observer is anxious to have as much of the southern sky as possible, assuming that he lives in the northern hemisphere (of course the reverse is true south of the equator), and it may be thought worth losing the north. However, the variable star observer or the general 'gazer' will want to have all directions available, and will have to compromise. The choice of a larger, permanently-mounted telescope or a smaller portable one will depend upon how good or bad the horizon is.

For observing the Sun, a refractor is far better than a reflector, and a 4in is ideal. The variable star or general stellar enthusiast will need wide field and good light-grasp rather than high magnification; the Solar System student will be more interested in magnification. In either case a 6in reflector (or larger) has much to recommend it. It will cost much less than a refractor of comparable power. Inch for inch, the mirror is less effective than the object-glass; thus a 3in refractor is a useful telescope, whereas a normal 3in reflector is not really adequate.

When buying a telescope, it is essential to be cautious. It is regrettable, but true, that there are plenty of bad telescopes on the market. In well-illustrated catalogues they look imposing, particularly if backed up by skilful sales-talk; but the only safe method is either to insist on a personal trial or else call in a trusted and knowledgeable friend to give an opinion. Fortunately there are plenty of good instruments available too.

Particular care must be taken when purchasing a reflector. A bad object-glass betrays itself more easily than a bad mirror, and the mounting of a refractor also tends to be easier to judge.

Few amateurs can make their own lenses, but mirror-grinding is another matter, and anyone who is reasonably skilful can do it, bearing in mind that the process is a laborious and rather messy one—and that things are always liable to go wrong, thereby ruining many hours of work. A couple of blanks and various essential pieces of simple equipment provide a start. The mounting is a sheer problem of carpentry or metalwork; the only other things that need to be bought are the flat and eye-pieces. It is usually wise to make a modest beginning, say with a 6in mirror, and experiment; it is probably true to say that one's second mirror is always far better than one's first, so that it is better not to be over-ambitious and start off with a 12in or even an 8½in mirror—though of course there are exceptions to this rule.

Therefore, I would suggest that the newcomer to astronomy should take the following steps:

1. Do some elementary reading, and learn the main principles.
2. Buy a set of simple star-maps, and learn the chief constellations.
3. Obtain a pair of binoculars, and explore the night sky with them.
4. Pause to take stock. Are you prepared to go to the expense, trouble or both of obtaining a telescope? If so, are you going to buy it or make it?
5. If the former, save for a 3in refractor, a 6in reflector or larger—assuming that you live where conditions for observation are not absolutely impossible (such as the middle floor of a flat in a large town).
6. Consult various catalogues, and, if possible, people who are experienced. Decide what your main interests are. If you are concerned mainly with the Sun, buy a refractor if possible—and bear in mind that the Sun should *never* be observed direct; blindness would undoubtedly result. If

you are interested more in the Moon, planets or stars, either a refractor or a reflector will serve, though the latter is apt to be cheaper.

7. If you propose to make a telescope, buy a couple of 6in blanks and the necessary abrasives, and begin work, bearing in mind that you may have to endure many disappointments before making a good mirror. Mount it, and test it out.

8. Join a society.

By this time you will have passed the novice stage, and you will be able to decide whether you want to carry on as a serious amateur astronomer. I hope you do—and that this book will be of some help in your decision!

Part I
TELESCOPES

2 : IAIN NICOLSON

Refracting Telescopes

AS IS WELL KNOWN, THE FIRST SERIOUS ASTRONOMICAL OBSERVA-
tions to be made with the refractor were carried out by Galileo
Galilei in the years following 1609, and the invention of the
instrument is generally ascribed to the Dutch spectacle maker,
Hans Lippershey, around 1608. The surprising thing about the
refractor—in principle at any rate—is its sheer simplicity, and—
because of this and the fact that spectacle lenses have been in use
since at least the thirteenth century—it is more than a little
surprising that the instrument was not discovered long before it
is generally reckoned to have been. There have been suggestions
that the principle was known before Lippershey's time, but
there do not seem to be records to support this contention.

The refractor predated the earliest reflector by at least sixty
years, but in its simplest form was severely restricted by false
colour effects—chromatic aberration, of which more later—and
the quality of available glass. Despite the fact that these problems have been to a large extent overcome, most serious amateur
astronomers today use reflectors, except in the smallest size of
telescope. There are perhaps several reasons for this, but the
prime objection to the refractor from the amateur's point of
view is the very high cost of a refractor compared with a reflector
of the same aperture. However, there are advantages and disadvantages to both types of telescope, and we shall see that there
are certain spheres in which the refractor scores quite heavily
over its rival.

To consider the refractor in more detail, it is worth saying a little about the process of refraction and the action of a lens. As shown in Fig 1, when a ray of light passes from a less dense

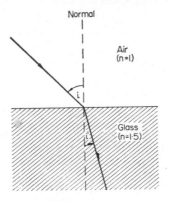

Fig 1   *Refraction at an interface*

medium (such as air) to a denser one (such as glass), the ray is bent, or refracted, in such a way that it emerges from the interface in a direction which lies closer to the perpendicular (or normal) to that interface. The angle which the incident ray makes with the normal (i) is the angle of incidence, and the angle after passing through the interface (i'), the angle of refraction. The amount of refraction that takes place depends on the properties of the media, characterised by the refractive index (n). A vacuum is taken to have refractive index $n = 1$ (and air, for practical purposes, is very nearly 1), whereas water has a refractive index of about 1·33, and most forms of glass have refractive indices of the order of $n = 1·5$. The amount of refraction which occurs when a ray passes from the first medium (n_1) to the second (n_2) is easily found from the formula $n_1 \sin i = n_2 \sin i'$.

It is seen from this formula that if a ray enters a piece of glass perpendicularly to the surface (ie when $i = 0°$), then it will be undeviated (for $\sin i = 0$, and therefore i' must also be 0°).

Bearing this in mind, we can look at the action of a lens. A lens has two surfaces, both or one of which may be curved.

Various forms of lens are in use in astronomical telescopes, and some of these are illustrated in Fig 2.

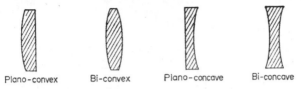

Plano-convex   Bi-convex   Plano-concave   Bi-concave

Fig 2   *Some common lens forms*

Let us consider the way in which a common biconvex lens produces an image (Fig 3). For this lens, the point *F'* is the focal

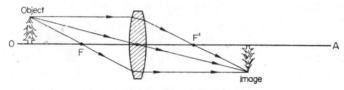

Fig 3   *Image formation by a lens*

point, that is, the point at which rays of light parallel to the optic axis (*OA*, which passes perpendicularly through the centre of the lens) are brought to a focus. The distance between the lens and *F'* is known as the focal length, and the ratio of the focal length to the diameter of the lens is known as the focal ratio. The point *F* (the front focus of the lens) has the property that a ray passing through this point will emerge from the lens in a direction parallel to the optic axis.

Let us now see what happens to rays of light coming from the object at *O*. Rays passing through the centre of the lens will carry on undeviated, while rays from the object parallel to *OA* will be refracted so that they pass through *F'*. Rays which pass through *F* will emerge parallel to *OA*. Where the rays meet up again, an inverted image of the object will be formed. If the object were to lie at an infinite distance away from the lens, then the image would be formed at *F'*. The closer the object, the further behind F' the image lies. Astronomical objects are so far away that they can be regarded as lying at infinity, and so the image is formed at the focus of the lens.

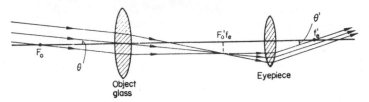

Fig 4 *The simple astronomical refractor*

As shown in Fig 4, an astronomical telescope in its simplest form consists of two lenses: a large lens of long focal length, known as the object glass (O.G.), which collects light from the object and brings it to a focus; and a lens of short focal length, known as the eyepiece, which contributes most to the overall magnification. If we consider a distant astronomical object which, when seen by the eye, has a certain angular size θ, then the O.G. will form an image of this (the prime focus image) at F_e'. If, now, the eyepiece is placed so that its front focus point f_e is at F', then rays of light from the image will emerge from the eyepiece at an angle θ', and can then be accepted by the eye. The magnification of the instrument is the ratio of the apparent size of the object with and without the telescope, ie, the ratio θ'/θ. It can easily be shown that this is equivalent to the well-known formula for magnification, M,

$$M = \frac{\text{Focal length of object glass}}{\text{Focal length of eyepiece}}$$

The type of telescope we have been discussing is the simple astronomical refractor described by Kepler in about the year 1611, and produces an inverted image. Galileo employed a concave eyepiece, giving an erect image, but his system had the disadvantages of a very small field of view. It is, however, still used in opera glasses and some forms of night binocular.

The severe problem which arises with the simple refractor is chromatic aberration (Fig 5). The amount of refraction which takes place in a lens depends on the wavelength of light, with the result that lenses act rather like prisms. Short wavelength light is refracted more than long wavelength light, so that blue light, for example, is brought to a focus closer to the lens than red light, and objects are thus seen to be surrounded by false coloured fringes. Resolution is consequently hampered.

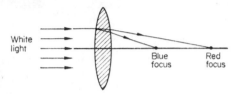

White light

Blue focus

Red focus

Fig 5   *Chromatic aberration*

The situation is mitigated if very long focal length O.Gs. are used, and in the seventeenth century in particular, this led to some extremely cumbersome telescopes being produced with focal lengths in the neighbourhood of 150ft and more (Huyghens employed a refractor of 210ft focal length). In such instruments the object glass was attached to a tall mast, and naturally, they were extremely difficult to use.

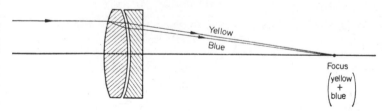

Yellow

Blue

Focus
(yellow
+
blue)

Fig 6   *Action of an achromatic lens*

However, in the eighteenth century, an English amateur, Chester Moor Hall, produced an O.G. made of two adjacent lenses of different types of glass which had greatly reduced chromatic aberration. The optician Dollond is generally credited with making serious use of this principle to produce good achromatic refractors. Traditional objectives of this type employed a component made of crown glass ($n = 1·52$) together with another composed of flint glass ($n = 1·63$), and in this way it was possible to bring two colours, at any rate, to the same focus (Fig 6). An O.G. which brings three colours to the one focus position is called an apochromat and usually consists of three elements. Fig 7 illustrates the focal points for different colours for a hypothetical achromat and apochromat.

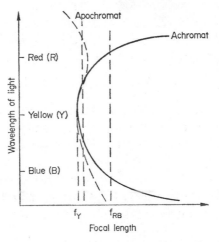

Fig 7　*Secondary spectrum in object lens—the dispersion in focal length with colour*

If two or three colours are brought to the same focus, then the other colours will be focused either in front of or behind the principal focus, giving rise to a minor spread of colour known as a secondary spectrum. Clearly, the important thing with an O.G. is not necessarily to bring two particular colours to the same focus, but to minimise the secondary spectrum—ie to reduce the range of focal lengths for the various colours. In other words, the curves of Fig 7 should be as flat as possible. For visual work, an achromat will normally suffice, provided that the curve is nearly flat in the yellow-green part of the spectrum, the region in which the eye is normally most sensitive.

In the early days of astronomical photography, plates were blue-sensitive and lenses were achromatised so that blue and yellow ('photographic' and 'visual') foci were the same—so giving rise to the photovisual refractor. With present-day panchromatic emulsions, apochromatic O.Gs. are more or less obligatory for many types of photographic work.

The amount of dispersion of colour foci with present-day optics is not great. G. D. Roth in *The Amateur Astronomer and His Telescope* quotes figures for a conventional two-element achromatic O.G. which has a secondary spectrum accounting for only about 0·3% of the total focal length. With really good three-element apochromats, the spread will cover only 0·05% of the total focal length. Needless to say, O.Gs. to the standard of the latter are extremely expensive, but the amateur will be perfectly well served for visual work, and many kinds of photographic work, by a straightforward two-element achromat.

Even with achromatic O.Gs., focal ratios are usually kept fairly high and f/15 is a fairly standard practice (in this case a 6in refractor will have a focal length of some 7½ft) and rarely is an instrument found less than f/12. At the other end of the scale f/20 is sometimes encountered.

So far, we have only considered chromatic aberration, but the refractor, like any optical instrument, is subject to several geometric aberrations. However, since the reflector is also similarly afflicted, it will suffice to point out that (i) telescope design is of necessity a matter of compromise—balancing one minor advantage against a complementary disadvantage, and (ii) the two-element achromat allows greater control to be exercised over these aberrations than in the case of the simple lens.

At this point it is worth listing the pros and cons of refractor and reflector. Firstly, as we have seen, the colour correction of a refractor is by no means perfect (although perfectly adequate for most amateur purposes); however, the reflector scores heavily here, as the angles at which rays of light are reflected are completely independent of wavelength, so that all colours are brought to the same focus. Secondly, the cost of a good achromatic objective is far in excess of a mirror of the same aperture, and so refractors are more expensive than reflectors of the same size. This necessarily follows from the fact that in the case of a mirror only one surface has to be ground, and the light does not have to pass through the glass (so that the optical quality of the glass need not be as good), whereas with the O.G. several surfaces have to be worked, and the quality of the glass must be very good. The long focal ratios of the refractor may tend to make the instruments rather cumbersome in the larger sizes (the 40in Yerkes refractor has a focal length of some 63ft) but there is some considerable compensation in that because of the long focal ratios (say f/15) the curvature of the optical surfaces is generally less than in the case of reflectors (which are typically f/8) and so the working of each surface is easier. For example, at f/15 one could get away with spherical surfaces, whereas for a reflector at f/8 a parabolic surface is essential.

Although these problems of cost and colour correction are obvious disadvantages of the refractor, none the less there are areas in which the refractor is superior. One important point is that, generally speaking, the refractor is a more powerful instrument per inch of aperture in terms of resolving power. This is particularly apparent in instruments of small aperture—a 3in refractor can be quite a useful instrument, while a 3in reflector is of limited value. The resolving power (R) of a telescope is generally estimated by means of Dawes' Rule, an empirical formula which states that $R = \dfrac{4 \cdot 5}{D}$ secs of arc, where D is the aperture of the instrument in inches. In other words, an instrument of aperture D should just be able to separate double stars with roughly equal components whose angular separation is R secs of arc. Now, in a perfect optical instrument, a star image should appear as a small bright disc surrounded by a series of faint concentric rings of light separated by dark spaces. The form of this—the Airy pattern—is indicated in Fig 8. For a

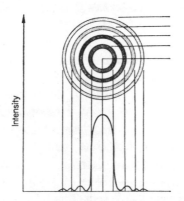

Fig 8 *The Airy pattern for a stellar image showing the telescopic appearance above and the intensity distribution below*

double star pair there will be two overlapping patterns, which will be just resolvable when the centres of these patterns are separated by approximately the distance given by Dawes' Rule. We can regard an extended image such as the Moon or a planet as being composed of a whole series of these Airy patterns, and so the ability of an instrument to resolve fine detail depends on the quality of the Airy patterns produced. Now a refractor, given perfect workmanship, should be able to produce ideal Airy patterns, but most forms of reflector, such as the Newtonian and the Cassegrain (see Fig 10), have secondary mirrors which interrupt the light path (and effectively obscure part of the primary mirror). It is qualitatively established, at any rate, that the presence of a secondary mirror will lead to a slight deterioration in resolving power; the effect is almost negligible with really small secondary mirrors, but increasing with increasing size of secondary relative to primary. Caessegrain reflectors tend to have larger secondaries than Newtonians, and so the effect can be more obvious. Furthermore, the supporting members for the secondary themselves lead to a certain amount of deterioration in resolving power. There are no hard and fast rules, but it is fairly safe to say that a 4in refractor is at least equivalent to a 6in reflector in this respect.

Considering amateur sized instruments, all forms of reflector (apart from the Herschelian, which has its own drawbacks) employ at least two reflections, and even with freshly deposited coatings, the light loss involved in the two reflections may be more than that involved in the passage of light through a good quality bloomed objective. However, mirror coatings deteriorate, and so for most of the time the light losses in a reflector may be considerably more than in a refractor of the same aperture. Apart from the problem of the maintenance of mirror coatings, the alignment of reflectors tends to need frequent adjustment while a refractor is more or less permanently aligned, and therefore in many ways more portable than a reflector.

It is well established that in poorer observing conditions, the refractor performs considerably better than the equivalent reflector. One of the principal reasons for this is the fact that the refractor, having the O.G. at one end of the tube, and the eyepiece at the other, is a completely closed system and therefore scarcely affected at all by the disruptive tube currents which occur in reflectors which are open at one end. (Although there is considerable debate as to the pros and cons of open lattice tubes versus solid tubes for reflectors, there does not seem to be any clear-cut advantage to either.)

A further point worth making is that the long focal ratios of refractors allow simpler eyepieces of longer focal length to be used to give the same magnification as reflectors of the same aperture. For example, a 6in f/16 refractor could achieve a magnification of × 192 with a $\frac{1}{2}$in eyepiece, whereas a typical f/8 reflector of 6in aperture would require a $\frac{1}{4}$in eyepiece (or a Barlow lens), which is rather short.

In summary one can safely say that in the size range of amateur instruments, a refractor will always out-perform a reflector of the same size, and in poor observing conditions will often prove superior to considerably larger reflectors. I have certainly encountered conditions where a 4½in refractor has proved superior to a 16in reflector. Of course, a larger instrument will always reveal fainter objects than a smaller one, and if, for example, one's interest lies in detecting faint variable stars, then a 10in or 12in reflector is clearly superior to a 6in refractor. Should one, however, be more interested in detecting fine detail, consistently, on a planet, then the reverse may be the case.

SUMMARY OF PROS AND CONS OF REFRACTORS

Advantages
1. Superior resolving power per inch of aperture
2. Superior performance in inferior conditions—image steadier
3. No reflections or interruption of light path
4. Near permanent optical alignment—minimum maintenance
5. Long focal ratios can mean use of longer focus, simpler, eyepieces

Drawbacks
1. Very high initial cost relative to reflector
2. A certain amount of secondary spectrum (chromatic aberration) unavoidable (reflector completely free of this)
3. Long focal ratios can mean that the instrument is cumbersome

The discussion so far has been limited to conventional refractors, but compound refractors are occasionally encountered, such as the Coudé, which has the advantage of a fixed eyepiece and a certain degree of compactness, although with the disadvantage of secondary reflections. If professionally made, Coudés tend to be enormously expensive. At the other end of the scale, the prismatic monocular (or binocular) is an example of a compound refractor. However, apart from the latter, compound refractors are rarely seen in amateur hands.

Should an amateur decide that a refractor would suit his purposes—and can afford the considerable cost—he will then find himself faced with the problem of actually buying one. If a small instrument of, say, 3in aperture is desired, then there are a considerable number of optical companies who make instruments of good quality in this size range. However, larger instruments do pose problems. As far as I am aware, the largest refractor in regular production in the UK at present has an aperture of 4½in (at a cost, fully mounted, in the region of £200 ($520)—comparable to the cost of a professionally made 8in reflector). In the United States one can find in the pages of *Sky and Telescope* instruments up to 6in aperture. However, for refractors in excess of 6in aperture, one moves into the realm of one-off instruments with a considerable increase in price. It is possible for the amateur to purchase achromatic objectives up to an aperture of about 6in and construct his own instrument without undue difficulty, but, whereas even a ham-fisted amateur can—with persistence—grind a mirror of 6in to 10in aperture, there are few amateurs who can undertake the grinding and assembly of an achromatic objective. The inherent cost differential between refractor and reflector is thus, in a sense, amplified further for the average amateur who, without considerable patience and application, would not be able to reduce his costs by grinding his own optics.

The other possibility is to purchase a second-hand instrument. Up to 4in or possibly 5in aperture, such instruments can be found without too much difficulty; beyond this size refractors seldom come on the market. It goes without saying that no second-hand instrument should be bought without being thoroughly tested and without seeking skilled advice, and this applies even more to refractors of 6in or over, where sums of several hundred pounds are involved. There is no substitute for testing an instrument under observing conditions, but these conditions must be borne in mind when assessing the instrument. Finally, an instrument with disappointing performance should not be rejected out of hand without first checking the alignment of the optical components, as it is always possible that the objective has been dismantled and reassembled with the components back to front!

Assuming the amateur has acquired a refractor, suitably mounted (the German mounting is most often encountered, although the modified English mount or the Fork have their advantages), which fields of astronomy are the most appropriate? From the foregoing it should be apparent that the refractor is to be preferred where consistent resolving power is of more importance than sheer light grasp. There are, perhaps, three areas of amateur interest where the refractor finds particular favour:

1. Visual observations of the Moon and planets.
2. Position angle and separation measurements of double stars.
3. Observations of sunspots, for most regular amateur observers tend to use refractors, including notable experts such as the late W. M. Baxter.

A further area where the refractor is pre-eminent, because of the permanence of the optical alignment, is in transit observations, but these are unlikely to be of interest to the amateur astronomer.

In conclusion, the refractor has much to commend it to the visual observer, but the bugbear of cost is a considerable one. The astronomer who decides to buy a refractor must bear in mind that for the price of his 4in refractor, he could have had a 6in to 8in reflector, for his 6in refractor a 10in to 12in reflector, for his 8in refractor a 14in to 16in reflector, and so on. Whereas 95% of the time, the refractor will resolve at least as well as the reflector in the same price range, there will be the odd occasion where conditions will be good enough for the larger reflector to approach its theoretical superior resolving power, and—of course—the reflector will have the fainter magnitude threshold. However, a large refractor is a very satisfying instrument, and fairly maintenance-free, compared to the somewhat more temperamental reflector. Personally, if I were looking for an instrument for serious visual observations and for the sheer pleasure of observing, then there is little doubt that I would go for a large refractor. Given average observing conditions, an aperture of 10in would be about the practical limit, but such an instrument would in many ways be a dream.

Reflecting Telescopes

THE TELESCOPE IS ESSENTIALLY AN INSTRUMENT FOR COLLECTING more light than would normally enter the eye. The average aperture of the eye is 0·2–0·3in diameter and only a small amount of light is available for vision; the smallest telescope has an aperture far exceeding this, and thus enables objects far too dim for normal vision to be seen.

In the small to moderate aperture telescope, the refractor is the type in general use, and uses a transparent lens to converge the light to a focus, or image point, as already described in Chapter 1. This focal point occurs at a distance behind the lens depending on the lens curves, the lens itself being composed of two components, one positive, one negative, the negative lens correcting the colour produced by the positive. The lens thus has four surfaces. The distance from the lens (the objective) to the focus is the focal length. If the focal length is 10in, it will yield an image at the focal point the same size as the unaided eye, this 10in being the accepted unit of closest distinct vision for a normal eye; if 20in, the image will be × 2, and so on. Subsequent magnification by an eyepiece results in the most generally known property of the telescope, that of magnification, though this is really the least important of its powers. The lens, needing four surfaces, and glass of extreme purity and precise qualities, is an expensive item, and is particularly expensive above 4in in diameter. Refractors of aperture 6in and over are rather rare, and the limit for this type of telescope was reached with the telescope of Yerkes Observatory, of aperture 40in. A lens of 49in aperture was made for a big exhibition in Paris in 1902, but was never successful.

It is fortunate that light can be manipulated by reflection as well as refraction, with a focal point in front of a curved mirror, instead of behind the lens as in the refractor. It is free from false colour, as the light remains intact during reflection.

The mirror of a reflecting telescope uses its leading surface only, the light never penetrating its substance; the clarity and purity of the material are therefore of no consequence. The mirror has to provide a bright mirror surface, and must keep its shape. Originally speculum metal, an alloy of copper and tin, whitened with arsenic, was used, as this metal could be polished to a brightness equal to a modern looking-glass, and was rigid enough. It was very heavy, as brittle as glass, and its polished surface was perforce part of its substance. As the mirror has to be correctly shaped to do its job, any repolishing to remove tarnish destroys the shape—necessitating reshaping, or 're-figuring'. This handicap persisted for over a century, during which period some exceedingly fine and, even by present-day standards, large telescopes were built, practically all in England, the home of the reflecting telescope. For a long time the largest telescope in existence, aperture 72in, was operating in Ireland, at Birr Castle. Built by the Earl of Rosse, it had a mirror (or rather mirrors, for there were at least two) of speculum metal. One of these mirrors is now displayed in the Science Museum, London, as is Herschel's 49in mirror, also in speculum metal.

The need for two mirrors for the 72in illustrates the drawback of metal as a mirror material. 'One on, one in the wash' had to be the motto. While one mirror was in use in the telescope (and tarnishing) the other was being remade, for that is what re-polishing means. Therefore quite conflicting reports of optical quality on the same instrument were equally valid, depending whether the observer had a session when the mirrors were in good shape or not. This major drawback to the reflecting tele-scope vanished when the method of coating glass with a thin film of silver was developed. Glass is nearly ideal for the substance of a mirror, being lighter than metal, less prone to changes of shape with temperature, and easier to bring to a perfect polish. It is interesting to note that speculum metal was so hard and brittle that it could not be cut or polished as metals usually are, but had to be ground and polished as if it had been glass. Therefore, no great changes in technique were needed. Silvering enabled glass to become reflective enough, as bare glass only reflects 4% of the incident light. The silver layer is the actual reflector, being so thin that it takes the shape of the glass without disturbing the accuracy of that shape. When tarnished, it can be dissolved off and a new coat put on quite easily and inexpensively. At once the reflecting telescope became the telescope of the future, for its optics could now be permanent.

Today, approximately a century after the development of silvering, the largest usable reflecting telescope has an aperture five times that of the comparable refractor. The silvering process is itself obsolete, aluminium being used instead as it can be kept in a serviceable state far longer than silver.

The main reason why the mirror has largely taken over from the lens is fundamental, not optical. The quest is always for more light, to see fainter and more distant objects, and therefore larger diameter light-collectors are needed, either lenses or mirrors. Each increase in diameter of a lens increases its thickness and its loss of light by absorption. The increasing difficulty of making larger glasses pure enough is nearly as insoluble. Light is also lost when reflected, but this loss is constant and independent of size of mirror. The glass for a mirror need not be pure; only mechanical conditions need be met, and size limits are financial, not physical. Glass itself is now being evolved into a substance that is not affected by temperature, and present-day mirrors for the giant telescopes can be immune to this complication.

The reflector's construction has changed greatly since Newton demonstrated that a curved mirror could do the work of a lens without the chromatic aberration inherent in the lenses of his day. His design is still in use, and bears his name. He took a curved concave mirror, placed it at the bottom of a tube and diverted the image out to the side of the tube by means of a small flat mirror placed at 45° in the open end of the tube. The diversion was necessary to avoid the observer's head obstructing the incoming light. (In today's giant telescope it is more economical of light to put the observer himself in a small room up in the tube mouth, instead of using a flat mirror.) Newton did not produce a telescope that worked well. The opticians of his day could not put the correct shape on the concave mirror, though there is no doubt that he recognised the figure needed. He was the pioneer, and others had to develop the means.

The reflecting telescope uses a mirror that receives incoming light from a far distant source, a star, practically speaking, at infinity, and hence with parallel rays. It must reflect it to one single point only, which is the image. A mirror whose section is part of a circle (a spherical mirror) cannot do this. The sphere can give a good image only if its object is at the centre of the sphere. The image then falls on it, and is perfect. Both object and image occupy the same focus, which is all that a sphere has. The parabola, on the other hand, has two foci, one at infinity and one at half its radius of curve (the basic spherical curve given to the mirror). An object situated at the distant focus will have the image formed at the nearer focus. This curve is achieved by deforming a previously spherical mirror in such a way that all rays falling on it from infinity come to the same focus. When finished, the mirror has had each concentric zone progressively changed in radius. Previously equal, they are now different from

centre to edge, the difference amounting to $\dfrac{r^2}{2R}$, with the centre

the shortest. In this, 'r' is the semi-diameter of the mirror, R is the radius of curvature of the mirror when spherical. The surface deformation applied to the original sphere is known as the 'correction'. When right, the mirror is said to be fully corrected. Under- and over-correction speak for themselves. For a Newtonian telescope intended for visual use (Fig 9), the focal length should be selected in the range f/10 to f/5, which means that the mirror is that number of diameters in focal length. Below f/5, the eye cannot accept all the light, and the higher powers can be reached only with powerful eyepieces, which are expensive, difficult to use, and involve problems of their own. Above f/10, the lower powers are lost unless special eyepieces are used, and the lower powers are the mainstay of observers. Reference to earlier comments on effects of focal length on image size will clarify this.

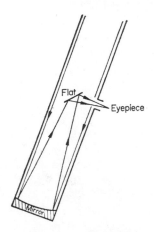

Fig 9 *The Newtonian reflector. Light passes down an open tube and strikes the main mirror. It is reflected back up the tube, and strikes a flat mirror placed at 45°. The rays are sent toward the side of the tube, and the image is magnified by an eyepiece*

If photography is contemplated, mirrors of f/3 to f/5 are possible with parabolic curves. The telescope has now become a camera. Should the telescope mirror be left uncorrected (spherical), and the necessary correction applied elsewhere in the system, f-ratios as fast as f/1 are reachable, with an extremely wide field and good quality images. This is the basic principle of the Schmidt camera (Fig 11), and the family of alternative designs that stems from it.

These instruments, the simple Newtonian telescope and the mirror/lens telescope, represent the development of the reflector to date. Between them come modifications of Newton's original concept, devised to make the reflected image more accessible. In a small or moderate-sized telescope, the position of the eye-piece, on the side of the highest end of the tube, is not too difficult to reach. With Newtonian telescopes larger than 18in aperture, the observing position can become almost inaccessible. Observatory instruments are mostly much larger than this, and some means of bringing the focus to a more manageable position is needed.

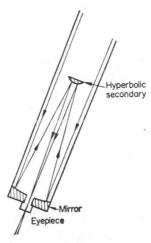

Fig 10 *The Cassegrain reflector. The secondary is hyperbolic, and the light is reflected back through a hole in the centre of the main mirror. A modification may be used; the light is caught by another small mirror before it reaches the main mirror, after reflection from the hyperbolic secondary. The light-rays are then deflected to the side of the tube, near the bottom end. At the expense of an extra reflection (and therefore extra loss of light) it has become unnecessary to make a hole in the main mirror*

It had been suggested by Gregory that the prime image of the paraboloid could be transferred to the bottom of the telescope tube by using a suitable concave mirror placed outside the prime focus. This secondary mirror has its surface figured until its section is that of an ellipse, which has two foci. An object placed in one focus will have its image produced at the other. Choosing the correct focus means that one focus will be close to the mirror, and the other at any specified distance from it. It is therefore simple to relay the prime image back down the tube and out through a hole cut in the centre of the main mirror. This central area is always 'dead ground', contributing nothing to the image, as it is directly shadowed by the secondary mirror and therefore receives no light. The secondary image is amplified by the secondary mirror, and is erect; all other types of reflectors give inverted images. The elliptical secondary mirror is quite simple to make, and the Gregorian telescope was soon in general use for apertures up to around 9in. The Gregorian was the first of the compound reflectors. Its main drawback was not serious for small telescopes—that since the secondary mirror is beyond the prime focus of the main mirror, the telescope tube has to be inconveniently long.

The logical solution was to change the secondary mirror from concave to convex, and put it inside the focus (see Fig 10). The figure for the secondary then had to be changed from an ellipse to its geometrical opposite, the hyperbola. Cassegrain, the inventor, was content to suggest it, nearly a century before the first convex hyperbolic secondary was actually made. However, reflectors built upon this pattern are still known as Cassegrains.

Cassegrains are the shortest of the reflectors, and the principle is used in one way or another in all the giant telescopes of today when long focal lengths in short tubes are needed. The saving in tube-length influences the sizes of observatory domes, to say nothing of engineering problems; it means that available finances can provide a telescope of greater aperture than would otherwise be possible. Optical efficiency in compound designs is no better than with the equivalent single paraboloid; but with really large telescopes the single parabolic mirror is impractical, and indeed unmanageable. Consider the 'hardware' which would be needed to cope with, say, a 100in mirror at f/50! The costs involved would be astronomical in every sense of the term.

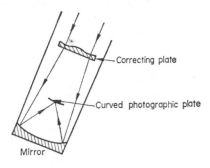

Fig 11   *The Schmidt, using a correcting plate and a spherical main mirror. This form can be used only photographically, with a specially-curved photographic plate, and is seldom found in amateur hands*

Until 1930, these three systems—Newtonian, Gregorian and Cassegrain—held the field. At this time, however, an Estonian optician produced a telescope with a spherical primary mirror. Such a mirror has no correction, and cannot produce a good image unless it is of hopelessly long radius. The telescope produced by Bernhard Schmidt had a correcting plate placed at the radius point of the spherical mirror, so that the correction was applied on the incoming light instead of being performed by the mirror surface. Such an arrangement means that all the incoming light has exactly the same correction, irrespective of its angle of entry, as all of it goes through the same entrance pupil. Since the mirror is spherical, all parts of its surface are identical, and deal equally with the already-corrected light, producing a field in which the image-quality deteriorates only very slightly toward the edge. This is a great improvement on the paraboloid, which produces perfect images on axis, but much poorer images away from axis. This is a severe limitation on the giant modern reflectors, whose f/ratios are extremely low.

As we have seen, the correcting plate in a Schmidt telescope is placed at double the focal distance in front of the main mirror. The telescope must therefore be double the usual length. Designs in which this disadvantage is removed have appeared since. In the Maksutov pattern, for instance, the correcting plate is placed at or inside the prime focus. This shortening of the telescope is produced at the expense of the correcting plate itself, which has to be extremely thick, and violently curved on both surfaces. Both the Schmidt and the Maksutov principles have been applied to the compound reflector, resulting in Schmidt and Maksutov Cassegrains. I have never been clear about the precise advantages of this development, as compounding means amplification. The original Schmidt-type image is of wide angular and small linear dimensions, and of overall excellence. It is just this wide, excellent field which makes the Schmidt so valuable, and at the prime focus the whole light-grasp can be utilised. Compounding, with its associated amplification, results in an enlarged secondary image which has to emerge through the hole (or its optical equivalent) in the main mirror before it can be used. Only the central part can do so, the rest being lost. But the virtue of the Schmidt lies in the marginal regions of the field! The field of a normal Cassegrain, or its equivalent long-focus paraboloid, is very good over half a degree or so—for practical purposes as good as that part of the field of the compound Schmidt which can be used. So why introduce an expensive correcting plate merely to give similar results? Nothing can equal the Schmidt and its derived types for prime-focus work. In this respect it stands at the peak of reflector development. Incidentally, it is startling to find that Kellner, known mainly for his wide-field eyepiece design, actually patented the Schmidt principle as early as 1910, twenty years before Schmidt himself came on the scene. Kellner's patent application and his drawings exactly define all the characteristics of the Schmidt, though he does not seem to have made an actual telescope on this design. One is reminded of the story of the first achromatic object-glasses!

In its various forms, the reflecting telescope can serve all fields of observational astronomy. Positional astronomy, formerly the unchallenged preserve of the refractor, is now the work of the Schmidt and the photographic plate. The enormous sky-coverage of the Schmidt makes comprehensive mapping possible in reasonable periods. Previously, with telescopes of large aperture but small field, such mapping programmes took decades.

Really big telescopes come into their own for more detailed investigations of specific objects. Such studies are rarely carried out visually. It has even been claimed that the era of the visual observer has passed. Certainly it is true that for astrophysical search, the spectroscope and the photographic plate, fed from the big mirror of a giant telescope, provide information quite beyond the power of the human eye. Very faint objects, at the limits of perception, can be reached only with enormous apertures. While the problems of the far reaches of the universe monopolise the largest reflectors, the bulk of astronomical research is carried out with telescopes of more moderate dimensions, only a few of which are refractors. The original drawback of the reflector—its inaccessible viewing-point—was solved by compounding the instrument, giving us the Cassegrain and the Gregorian. The Gregorian design is rarely used, for reasons which will be clear from what has been said above; but both systems give an image which can be put anywhere by choosing a suitable curve for the secondary mirror. The original embarrassment is now an advantage, for the secondary image can be led underground and fed into a variety of auxiliary instruments. Such underground laboratories are now incorporated in many observatories, particularly those concerned with studies of the Sun.

The reflector has emerged as the tool of the professional

astronomer; it has always been a favourite with the serious amateur. In sizes up to 8in aperture, the properly-figured reflector is only slightly less efficient in light-grasp to a refractor of corresponding aperture; it is its equal in resolving power, and is capable of better performance. At 8in aperture, it begins to excel the refractor in light-grasp. It is always much less costly than a refractor, and is therefore more commonly found in the hands of non-professional astronomers. In smaller sizes, from the 1in aperture of small binocular field-glasses through the range of spyglasses and terrestrial portable telescopes up to the 3in or so of coastguard and naval telescopes, telescopes are normally of the refracting type. The refractor's superior robustness, closed optics (lenses at each end of a metal tube) and freedom from adjustment worries make this end of the aperture-scale its exclusive province. Refracting telescopes are common up to 6in aperture, and several 8½in and at least one 10in are in use by modern amateurs. However, almost all serious amateurs use reflectors, because of the extra aperture available.

Throughout this survey, aperture has been quoted as the most desirable feature of any telescope. Perhaps it is time to clarify this statement. To do so, the actual function of the lens or mirror must be examined. Consider light emanating from one single source at an enormous distance—for example, a star, which is virtually a geometrical point seen at infinity. The light reaches us in a parallel stream. The object-glass or mirror (considered here as being beyond reproach!) accepts this parallel stream, and turns it into a converging cone of light. This cone converges, crosses and then diverges again. The crossing of the light is the focal point. But, notwithstanding that the source is a geometrical point, the focal condensation is not. It is a tiny disk containing most of the light. The rest of the light surrounds this disk in rings of diminishing brightness. So we have one point—the object—producing one disk—the image.

The actual size of the focal disk is inversely proportional to the diameter (i.e. aperture) of the objective which produced it. Therefore, the larger the aperture used, the smaller the focal disk. Consider now the function of separating, or resolving, adjacent objects, which may be expressed as $R = \dfrac{4 \cdot 56 \text{ seconds of arc.}}{\text{aperture in inches}}$ If we now imagine that there is another star beside the first (at infinity, of course), another image will be produced beside the original in the focal plane. If the second star is exactly 1 second of arc away from the first, we will find that no telescope below $4\frac{1}{2}$in aperture will show the two stars as two separate images, because the separation of 1 second of arc is beyond the resolving power of smaller apertures. An aperature of 9in is needed to resolve objects of separation of a $\frac{1}{2}$ second.

The resolving power is thus a direct measure of the smallest possible detail which can be present in the image. A telescope image can be visualised as being composed of separate individual units, rather in the manner of a mosaic pavement. In the picture or pattern of a mosaic pavement, it is impossible to show any detail which is smaller than one of the units composing it. If such detail must be shown, it is necessary to use a smaller tile or unit. The case of a telescopic image is similar. If finer detail is needed, then a larger aperture must be used. This has nothing to do with the magnifying power of the telescope. A power of $\times$ 200 on a $4\frac{1}{2}$in telescope gives an image of exactly the same size as $\times$ 200 on a 45in aperture telescope: both are $\times$ 200. But the image with the $4\frac{1}{2}$in telescope can show no detail finer than 1 second of arc across, while the image of the 45in instrument can show detail down to 1/10 of a second of arc.

Sharpness and clarity of the image is termed 'definition'. This concerns the quality of the telescope, not its size. No increase of magnifying power can render visible detail which is not present in the image—which places mere magnification in its proper, very secondary perspective. One second of arc is the angle given by lines drawn to the two sides of an object 1 mile across lying at a distance of 206,265 miles away. This is approximately one part in 200,000, and is also, approximately, a crater 1 mile across at the distance of the Moon. To be able to see such a crater requires $4\frac{1}{2}$in of aperture. If $\frac{1}{2}$-mile craters are to be seen, 9in aperture must be used; for $\frac{1}{4}$-mile craters, 18in aperture. Most astronomical objects are much more distant then the Moon, and their images are correspondingly smaller, so that the importance of resolving power cannot be over-stressed.

The second important factor is light-grasp, which is related to resolving power. Objects which are sufficiently far away can be too dim to be seen. Increased aperture collects more light to put into the image, and here the reflector has the advantage over the refractor. Increasing the aperture in a lens means increased thickness, and therefore increased light-loss because of absorption in the glass. In refractors, unit transmission is therefore a diminishing factor. With mirrors this is not the case. Unit reflection is the same whatever the diameter of the mirror.

Because light in passing through a lens is split up into its component colours, and has to be recombined, a focal length of 15 diameters is needed to give the best quality result. This gives a $6\frac{1}{2}$in refractor a focal length of approximately 100in. A $6\frac{1}{2}$in mirror can also be of 100in focal length, but it does not have to be so. It can be made anything desired; but 8 diameters, giving 50in focal length, allows theoretical perfection to be reached without undue demands on the mirror-maker. It is therefore much shorter than the equivalent refractor, and has its main weight at the bottom of the tube—which makes the mechanics of the mounting much simpler and cuts down the cost.

Magnifying power, which is $\times 1$ for each 10in of focal length, gives $\times 10$ for 100in, but only $\times 5$ for 50in focal length. This is the primary image magnification, compared with the size of the image as seen with the naked eye. The application of magnifying eyepieces to this gives the final power of the telescope. Eyepieces are defined in terms of their own focal length. Their function is to enable the observer to inspect the primary image from distances much less than those normally needed by the eye (about 10in). An eyepiece of 1in focal length allows the observer to see the image clearly at 1/10 the normal distance, so that it will also be ten times larger. Used on the × 10 image at 100in focal length, this will give a magnification of 100. Used on the 50in focal length instrument, the magnification will be only 50. To obtain equal final magnification, the reflector will need eyepieces of nearly double the power (ie half the focal length) of those used on the refractor. This, together with the larger angled cone of the reflector, makes it desirable for reflector eyepieces to be completely free from serious errors of their own; imperfections are more important than with eyepieces used on refractors. While simple eyepieces (Huyghenian, Ramsden and Kellner) function perfectly on the refractor, something better than the Huyghenian, at least, is needed for the reflector. Achromatic eyepieces (orthoscopic or monocentric) are better than the simple types. It is not easy to obtain eyepieces of focal length much shorter than $\frac{1}{4}$in, and neither is it advisable. Apart from their cost, their restricted field of view, curvature of field and short eye-relief, make them troublesome to use. If the object under study demands a high power, a better way to get it is to increase the prime focal length, and hence the size of the primary image, by using a Barlow lens and a normal eyepiece. Generally, however, the image detail can be exhausted without the use of inordinately high powers. As has been said above, magnification is of secondary importance.

4 : T. W. RACKHAM

Principles of
Mirror-Making

IT SEEMS THAT THERE ARE TWO SEPARATE AND COMPULSIVE
influences that are able to affect a person to such a degree that
he (or she) can no longer resist the urge to own an astronomical
telescope. The first is an interest in astronomy supplemented by
the desire to see for oneself the wonders of the universe. The
second stems from the science of optics itself. Needless to say,
there are many branches of optics that are not remotely associated with astronomy, so I am going to assume that the reader,
having arrived at this point, has done so because he has an
interest in astronomy and telescopes. Or more specifically because he wishes to acquire and own an astronomical telescope.
If the latter is the case, then it becomes a question of cost, which
depends on the size of the instrument as well as its condition—
here I am including new and second-hand instruments. Astronomical telescopes, although not always as stable on their
mountings as they should be, seem to be remarkably stable in
price. Let me quote you from the Catalogue of Instruments
made by Troughton and Simms which is contained in a volume
entitled *The Achromatic Telescope* that was published by William
Simms in 1852. 'Item 40: Achromatic telescope, completely
mounted, equatorial, with clock movement, micrometer, etc.,
5 feet focus and 4-inch object glass . . . £230.' A quick glance
through the telescope advertisements at the present time will
reveal that the British price for a $4\frac{1}{2}$in refractor—all right, so we

have an extra $\frac{1}{2}$in of aperture and no micrometer; and an electric driving motor instead of the mechanical clock—is £275! Ancient and modern American prices, although somewhat higher than those in the sterling area, show a similar trend. Notwithstanding this, there is one important omission from the 1852 Catalogue of Instruments made by Troughton and Simms—no mention is made of a reflecting telescope and, indeed, the delicate Foucault Test, so necessary to gauge the perfection of a telescope mirror surface, does not seem to have been announced until 1858. It is true to say that this, coupled with a method of depositing metallic silver on glass, revolutionised the manufacture of reflecting astronomical telescopes.

Because there are four optically worked surfaces in an achromatic objective and only one in a telescope mirror, the price of a reflector is far less than that of a much smaller refractor. Returning to the modern advertisements, the £275 quoted above for a $4\frac{1}{2}$in refractor would buy a 10in reflector and provide some monetary change for the purchaser. In performance, the reflector far outweighs the refractor as far as image resolving power and 'light-grasp' are concerned. This means that the reflector enables its owner to 'split' closer double stars—that is, to separate two close stars that in a smaller telescope might appear as one—and to observe finer lunar and planetary detail. In fact, if observing conditions and the optics were perfect, the 10in reflector would possess image resolving properties that would be nearly $2\frac{1}{2}$ times better than those of the $4\frac{1}{2}$in refractor. Light grasp is directly proportional to the area of the objective or mirror, so, from this point of view, the 10in mirror would collect and focus nearly five times as much light as the $4\frac{1}{2}$in objective would for the same celestial object. Due to the presence of a secondary mirror, fixed above the primary, there is a loss of light in the reflector which reduces the 'light-grasp' ratio to about 4:1. This in turn allows the observer to double his magnifying power on the reflector and maintain the same brightness of image as would be obtained with the $4\frac{1}{2}$in. Perhaps an example might clarify this statement: for a given object, a power of 100 on the refractor would provide an image brightness that would be approximately matched by the image brightness in the reflector with an eyepiece magnifying 200 times. So, when it comes to making a choice from the several types of astronomical telescope that are available, most folk are concerned with value for money, and unless the chosen instrument is to be no more than an ornament to decorate an unused corner of a room, there is a persuasive argument in favour of the reflector. So far so good, but there are several types of reflectors which include the Newtonian, the Cassegrain, the Gregorian, the Schmidt and related cameras. We can quickly delete the Gregorian, with its concave secondary mirror, from our list since this is largely of historical significance. The Cassegrain has become very popular in recent years because of its ability to provide highly enlarged images of the planets and portions of the Moon. The focal ratio of this type of optical combination is large—f/15 to f/30 is common—and it is this property that prevents the Cassegrain from being used where larger angular fields have to be observed. For example, if the observer wishes to see the whole of the lunar image, or Jupiter with the Galilean satellites or the bright undiluted fields of star clusters such as the Pleiades, then the Newtonian reflector must be used, or at least a Cassegrain/Newtonian combination. In other words the Cassegrain convex secondary is exchanged for an inclined Newtonian flat mirror, and the focal ratio of the primary mirror—popularly f/6 to f/8—dictates the angular diameter of the field. Because the amount of the observed field depends on the diameter of the field lens in the ocular or eyepiece—this is the eyepiece lens furthest from the eye—it can be appreciated that this lens can accept a much smaller angular area of the field of the Cassegrain system.

With the threat of metrication in the air it is no less true to say that one cannot pour a litre of liquid into a half-litre pot, and the angular field of a Cassegrain telescope working at f/30 is dimensionally 25 times the area of the same angular field of the f/6 Newtonian. Assuming that our eyepiece field-lens is 1in in diameter, and that is accommodates a Newtonian angular field diameter of 1°, then, with the Cassegrain system, it can accept only 1/5° of the field, and the 'overhang' areas cannot be observed without adjusting the position of the telescope. It is important to note that this condition is true only where the primary mirrors are of equal physical diameter. The sizes of images formed in telescopes are proportional to the focal lengths of the mirrors or objectives used—double the focal length and the resultant image diameter will be physically doubled and its area quadrupled. Nothing that we can do can alter the angular diameter of the astronomical object itself: the Moon's angular diameter from Earth remains at about $\frac{1}{2}$° no matter how we alter mirror focal lengths or eyepiece magnifications.

The other useful property of the Cassegrain system is that the high magnifications are obtained with mirrors and not with lenses, so that the field of this type of telescope is virtually free of chromatic aberration. High magnifications can be obtained with the Newtonian system by using high-power eyepieces, and these can introduce chromatic effects that are noticeable around the edges of the field. Much depends on the type of eyepiece, used and it is probably true to say that the amount of perceptible residual colour observed is inversely proportional to the price paid for the eyepiece—there are modern multi-element eyepiece combinations that are almost free from colour defects.

So where does this get us? To avoid any confusion let me say that the purpose of this preamble is to outline briefly the advantages and disadvantages of various reflecting telescope systems, the main contenders being the Newtonian and Cassegrain. I think that enough has been said to show that the Newtonian is the best 'all-rounder' and that the Cassegrain is the ideal close-up telescope for the lunar and planetary specialist. This is not just my opinion: you will see more Newtonian telescopes advertised than other types, and since this is based on the down-to-earth 'supply-and-demand' tactics of the manufacturers, I believe that this very eloquently argues the case in favour of the Newtonian.

The easiest way to acquire a Newtonian reflecting telescope is to buy one if the price is no obstacle. If the price is an obstacle, even to the procurement of a second-hand instrument, then other means must be sought and the cheapest way to obtain a reflecting telescope is to make it oneself, starting with the all important telescope mirror. Most people imagine that the manufacture of astronomical telescope mirrors requires the use of expensive and sophisticated machinery which, of course, is true if one gets involved in the mass-production of telescopes. However, the most useful and versatile tools endowed to Man are his hands, and with these, and very few other inexpensive items, he can manufacture a telescope mirror that will afford him infinite pleasure, not only in the making, but also in the using. While there are yet to be found those intrepid and, maybe, masochistic individuals who prefer to cut out their mirror disks from sheets of plate glass, I personally believe that this is an unnecessary chore that sorely tries the patience. I would recommend that the aspiring mirror-maker should buy a 6 or 8in diameter Pyrex or Monax glass blank. He can, if he wishes, buy a blank that has been hollowed to the approximate concavity of the finished mirror surface, but because the grinding stages of mirror making are the easiest (and noisiest), I believe that the tyro should undertake these too. With smaller mirrors, and where weight is of no importance, the thickness of the mirror blank is usually 1/6th of the diameter, so a 6in mirror is initially 1in thick. While there are no hard and fast rules concerning the diameter/thickness ratio, there seems to be general agreement that it is dangerous to make the thickness less than 1/10th of the diameter, for the mirror may not contain sufficient glass to maintain the shape of its one optical surface. This brings us to the subject of what a telescope mirror is and to the essential principles of how such an object is made. We have discussed loosely the mirror disk or blank, but what we described is not so much the mirror as its support. A telescope mirror is a paraboloidal surface of silver or aluminium which is accurately figured to literally millionths of an inch, and it has to preserve this accuracy no matter which way it is turned or affected by the earth's gravitational field. Recognising the impossibility of manufacturing such a disembodied entity by itself, mirror makers take a thick blank or disk of glass and grind and polish one side of it until the desired perfection of figure is attained. Then, and only then, is the figured surface coated with a microscopic thickness of aluminium which becomes the telescope mirror.

The principles of mirror making are well known, but the literature on the subject was fragmentary and scattered until the Rev W. F. A. Ellison, one time Director of the Armagh Observatory, Northern Ireland, set a trend that has been followed by several other workers, and published his little book entitled *The Amateur's Telescope* in 1920. This book described how the aspiring mirror maker should set about the task, and listed the necessary items that he would require for the making of mirrors as well as for achromatic doublet objectives. While some of the grinding powders are no longer available, the methods described by Ellison remain untouched by the intervening years. The essence of mirror making depends on there being two circular disks of glass, the flat faces of which are ground together with abrasive powders. The amount of glass that has to be removed initially is large, and this has to be done by using large diameter particles of abrasive.

To understand the method of generating spherical or spheroidal glass surfaces, we have first to consider the mirror making tool. Where only one mirror has to be made, this takes the form of a glass disk the same diameter as the mirror blank. This is fixed centrally to a pedestal on a very solid base on, preferably, a stone or concrete ground floor. The pedestal is arranged so that the operator can walk completely round it while he makes his mirror. This also means that the height above ground of the upper horizontal flat surface of the glass tool has to be about the height of a normal workshop bench. The pedestal must be solid and inflexible and free from vibration. The surface that is destined to become the telescope mirror is worked face down on the upturned surface of the tool, and, as we have already observed, grinding powders are used to excavate the curved surfaces. The glass surfaces are worked together in a process that combines three distinct and separate movements of the operator. Let us assume that the mirror maker is grasping the mirror blank in both hands and is exerting a downward pressure. At the same time, he alternately pulls the blank towards him and then pushes it away from him. This movement is called the *stroke*. It is clear that the stroke provides varying amounts of overhang when the mirror blank is no longer concentric with the glass tool. Thus, if he commences with the centres of both disks together there will be no overhang. But, as the upper disk is pulled towards the operator, the overhang increases to a maximum at the position where the stroke is stopped as a prelude to it being commenced in the opposite direction to a second maximum, where the overhang takes place on the side of the tool furthest from the worker. The stroke is just as cyclic as the movement of a clock pendulum or of that of a piston in an internal combustion engine.

At first sight it seems that two flat surfaces, ground together with abrasive powders in the manner just described, ought to remain sensibly flat. Why, then, should the upper surface of the glass tool become convex? And why should the lower surface of the mirror blank become concave? The explanation lies in the fact that the two glass disks do not represent a symmetrical system. It is true that two concentrically placed glass disks appear to be symmetrical, but the effects of the considerable downward pressure of the operator and the weaker forces of gravitation disturb the system. As a consequence we have a situation in which the amount of glass removed from both surfaces depends on two entities: firstly, the rate at which glass is removed will depend on the downward pressure exerted by the operator. For the moment there is no need to consider the cutting power of the abrasive powders. Secondly, consideration must be given to the area of the glass surfaces actually in contact, since no cutting can take place where there is no contact, or, in other words, where there is overhang. Cutting action is at its maximum at the ends of the strokes where the overhang is also at a maximum, but it is, of course, confined to those smaller areas where contact exists between the two glass surfaces. Minimum grinding action occurs in that part of the stroke where the centre of the mirror disk passes over the centre of the glass tool. At this instant the downward pressure of the worker is uniformly distributed over the entire areas of both mirror and tool. Because it is not possible, and indeed not necessary, for the mirror maker to change his downward pressure for different positions of the stroke, we can see that maximum cutting, taking place in the extreme overhang positions of the stroke, removes glass from the sides of the tool and from the central regions of the lower surface of the mirror disk. Nevertheless, very peculiar shapes would be generated if only the stroke were used as described. In order to ensure that grinding takes place evenly and smoothly, on the lower surface of the mirror disk, the operator has to rotate the mirror blank slowly in his hands. The third motion, that ensures that the upper surface of the glass tool is evenly and symmetrically abraded, is performed by the operator slowly moving round the pedestal that supports the tool. While there are no hard and fast rules, the turning of the mirror blank in the hands should take about as long as the operator takes to circumnavigate the pedestal—between one and two minutes has been found to be about right. With these three movements, and a few different grades of abrasive grits, a mirror-maker can soon excavate a mirror blank to the desired depth. Cutting power is largely a function of grit size—it is not difficult to imagine what happens on the microscopic scale where an irregular grit fragment, sandwiched under pressure between the two glass surfaces, rolls under the action of the stroke and produces tiny conchoidal fractures in both glass surfaces. The glass particles mix with the abrasive, and the depressions left behind are called 'pits' and it is essential that smaller grades of abrasive powders be used to remove the pits left behind by the larger ones. Generally, the mirror surface can be excavated to its desired curvature with a large fast-cutting grit followed by a smaller slower-cutting grit. The finer abrasive powders that are used after this are required to produce a surface that will polish without leaving unsightly pits that hurt pride rather than the performances. These fine powders do little to deepen the curve of the mirror surface.

This, then, is an outline of the time-honoured method of generating spherical surfaces that was probably known in antiquity, almost certainly used by thirteenth-century spectacle makers, and very definitely employed by Newton and Herschel. For the moment there is no need to discuss specific topics such as focal lengths, focal ratios, and radii of curvatures, etc. Let us assume that the mirror maker has ground his surface to the desired depth and, that by using various other powders graded in diminishing sizes, has produced a finely ground surface that will take a polish. This finely ground surface will not reflect light specularly, but will appear very smooth and grey. No pits should be visible when this surface is observed under a powerful hand magnifier.

POLISHING

The technique of polishing glass mirror surfaces is essentially similar to that of grinding, but with one important difference— the glass surface of the grinding tool is replaced by a softer and more yielding surface of pitch. Also the polishing powders are finer and smaller than the grinding grits. Nevertheless, the division between polishing powders and grinding powders is not even hair-line, because a fine grinding grit polishes if it is used on a pitch lap and a polishing powder will grind if placed between two glass surfaces. It is the yielding surface of the pitch lap that traps the polishing powder particles and holds them in place and permits each to be adjusted to the height of its fellows. It seems that there can be no rolling action similar to that which occurs when two glass surfaces are used as in the grinding operation. Over the years there has been considerable argument as to what polishing really is: some have regarded it as a refined continuation of the grinding process while others have invoked theories involving molecular flow. None the less, while there may be disagreements among the experts, the fact remains that a finely ground glass surface will take on a superb polish if the three actions of grinding are continued with the use of a pitch lap and a mixture of rouge powder and water.

Disregarding the tricks of the trade used in the optical industry, and confining ourselves to a discussion of the simple methods that can be used by anyone, a pitch lap usually takes the form of a layer of pitch—or a mixture of pitch and other ingredients—that is made on top of the upturned and scrupulously cleaned convex surface of the glass tool. To reduce adhesion and suction between the lap and the mirror, and to improve the circulation of water and rouge mixture, and also to permit the pitch to flow more easily in moulding itself to the perfect shape of the finely ground mirror, the lap is cut into squares. The material between the squares is completely removed. Thus we have a black surface that, apart from its colour, shape and modest convexity, reminds one of the familiar 'easy-to-break' chocolate blocks. Cutting the lap in this way disturbs the surface so it is necessary to 'cold press' the lap for several hours so as to ensure that the lap flows to the perfect shape of the mirror once again. Then, and only then, is it safe to commence the polishing process. The worker uses the same three movements—the stroke, the turning of the mirror blank in the hands and the bodily movement about the pedestal—and if all is well the mirror surface should take on a fine polish after a few hours of work.

Having acquired a polished surface, it follows that this is a surface that will reflect light, and thus we can assess the quality of this surface by using the already mentioned Foucault knife-edge test. This optical test, which will be described in the next section, is extremely precise and, although it requires very simple apparatus, it can still measure imperfections on a mirror surface that amount to no more than a few millionths of an inch.

With the completion of the stages of grinding and polishing, the reader may be pardoned for thinking that this is the end of the mirror-making process and that the next step is to aluminise or silver the newly formed surface. And with a mirror of very long focal length this could be the case. Usually, however, the mirror surface is not left spheroidal, but has to be figured to the shape of a paraboloid. This involves a special form of polishing, the purpose of which is to deepen the curve of the central regions of the mirror. So little glass is removed that the pitch lap and rouge powder and water mixture can be used to effect this transformation. There are several standard techniques, all designed to increase the polishing efficiency in the centre of the mirror without significantly changing the shape of the peripheral regions. Whereas the grinding and initial polishing of a telescope mirror do not pose any insurmountable problems, the parabolising can introduce snags, and the newcomer may find that his mirror goes through all sorts of shapes before he finally triumphs over the obstinate glass.

Once the perfection, or near-perfection, of the mirror surface has been assessed by the delicate knife-edge test, the amateur mirror-maker has the satisfaction of knowing that all of the remaining tasks of making the tube, and mounting the telescope, are comparatively simple. Aluminising is a process that requires expensive vacuum pumps and associated apparatus, and it is best to entrust the coating of the mirror to laboratory experts or to optical firms that undertake this type of work.

Mirror-making for the beginner can be a lengthy task, but it is also a pursuit that can be divided into separate stages. For example, the rough grinding to the approximate curve could be completed in an evening, providing that all of the equipment is at hand. Fine grinding is not a long process, but because it is not easily possible to gauge the smoothness of the mirror surface or to observe the pits that are left from the larger grits, fine grinding is usually done 'by the clock'. In other words, the assumption is made that it is safe to continue with a given grit size for a certain length of time, depending on the size of mirror and the stamina of the operator. After this time has elapsed, it is safe to wash down the mirror and tool so as to remove all traces of that grit in preparation for the smaller size that has to follow. This is the golden rule all through the grinding process.

Polishing and parabolising are other stages that can be taken on piece-meal or as a continuous process, depending on the time available. Generally, it is convenient to follow the polishing by the parabolising without too much delay, because of the temporary nature of the pitch lap that can flow very considerably over the period of a few weeks.

TECHNIQUES

In the last section we discussed some reasons for choosing the Newtonian reflector and briefly reviewed the outline of the mechanics of mirror-making. We shall now consider in more detail those subjects that I consider to be basic and essential, starting at the point where the aspiring mirror-maker has received the package containing his mirror blank and glass tool (or any other special type of tool that is available), grinding grits, optical rouge powder or cerium oxide (or any other polishing powder disguised under its trade name). Let us assume also that he has found a suitable workshop—an underground cellar where the temperature remains sensibly constant is ideal—and in it he has installed the solid pedestal to support the grinding tool. For a pedestal, a cast concrete block can be made, or a little brick-laying will produce a suitable structure; a large barrel filled with bricks works well; a chimney-pot filled with a mixture of stones and concrete, with its cylindrical form, approaches the ideal. Failing these, a hunt through the local scrap-heap will unearth something that will serve this purpose. Whatever is used should be capped with a stout wooden board into which wood screws can be driven to fix the glass tool in position. There are several ways of doing this—one way is to cement the tool with pitch to a flat disk or hexagonal of wood of smaller diameter. The wooden disk is attached on its lower face to a wooden or metal base considerably larger than the grinding tool. Wood screws can be driven through the wooden or metal base into the wooden pedestal top. Three such screws, spaced at about 120° on a circle an inch or two larger in diameter than the grinding tool, complete the job—for obvious reasons the circle cannot be smaller in diameter than the glass tool.

It is convenient, but by no means essential, if the workshop contains a water supply and sink and an electrical heating ring or gas ring (though the former is the safer when one is trying to melt pitch and turpentine). Other requirements, old saucepans, spoons, soap, etc, are easily found around the house.

FOCAL LENGTHS AND RADII OF CURVATURE

The first thing that has to be decided is the focal length of the mirror—this and the diameter will have been decided beforehand by the purchaser if he has ordered a mirror blank that has been ground to curve. Usually the focal length of a Newtonian telescope mirror is about six to eight times the diameter of the blank; this is expressed as the focal ratio, a symbol that is familiar to every photographer. Because of the deeper curve and the greater difficulty with the parabolising, the beginner usually makes his first essay a f/8 mirror rather than a shorter focal length f/6. An f/10 is an easy mirror to make because the difference between the spheroid and the paraboloid is so slight that it can be ignored. Nevertheless, a 6in f/10 mirror has a focal length of 60in, which is on the long side when one comes to find a tube for such a thing.

When we come to the business of grinding out a mirror to curve, we start with the focal ratio, for this determines the focal length for a given mirror diameter. Thus, if we choose a 6in diameter blank and a focal ratio of 8 (f/8), our focal length will be eight times the diameter, or 48in. This is another way of saying that the images formed in front of the finished mirror, by objects situated at infinity, will be 48in in front of the parabolised surface. This is the focus, and this must not be confused with the actual radius of curvature of the mirror surface, which is twice as long. So, in the case of our 6in mirror, the surface will resemble a small circular area of a large sphere whose radius is 96in, and, in point of fact, this curvature provides a depth of about 0·1in—this being the depth of the depression in the centre of the mirror as measured from the edges.

GRINDING
The besetting problem during the early stages of the coarse grinding is to determine and control the amount of glass that is being removed from the mirror. It is the large grits that do most of the excavating. There are several standard techniques, some of them very simple—the late Rev W. F. A. Ellison wetted the newly ground mirror surface, thus creating a temporary mirror which would reflect light back to the vicinity of its source to reveal the radius of curvature. Or again, a sheet-metal template carefully shaped to the required curve could be made and used to gauge the depth of the concavity. But the most accurate way of telling how the grinding process is going is to measure the depth with a spherometer, which is a tool that stands on three equidistant metal feet and in the centre of which there is an accurate micrometer screw; or failing that, a direct reading dial-gauge. Place the instrument on an optical flat and it will read zero: place it on a concave mirror and it will read the depth between the three metal feet.

In the case of the grinding of a spherical concavity, the spherometer reading (*d*), the radius of curvature of the surface

(R), and the radius of the circle containing the three metal feet (r) are related thus:

$$d = \frac{r^2 - d^2}{2R}$$

For long-focus mirror work the d^2 term can be omitted because the value of d is not likely to be more than a few thousandths of an inch. It goes without saying that the mirror surface should be well sluiced with water to remove any abrasive grits before the spherometer is used.

Most of the excavation work will be done with silicon carbide grits that are disguised under a number of trade names. Fused alumina grits do not have the 'bite' of silicon carbide but, nevertheless, they will remove the unnecessary glass at a satisfactory rate. These coarse grits should not be used to excavate the mirror surface to the calculated depth, and we would do well to remember Ellison's advice that the rough grinding should stop when the length of the radius of curvature, R, is within 8–12in of the desired length. At this stage the spherometer is almost indispensable.

The mirror surface is brought closer to the desired depth by changing to a finer grade of grit, which has an average grit size approximately half that of the former grit. It should be understood that no manufacturer of grinding powders will guarantee that all of the particles of a certain grit size *are* the same size— instead he will tell you that they fall between certain dimensional limits. The user should also play safe by *never* having more than one grit container open on the work-bench at a time. And, again, it cannot be over-emphasised that the mirror blank and tool and the surrounding work area should be washed down thoroughly before changing from a coarser grit to a finer one. The presence of just one or two particles of the larger grit can cause unsightly scratches that may not be revealed until the finer stages of grinding have been completed. There are no hard and fast rules concerning the number of grits that should be used during the grinding stages. Ellison used one for roughing out, and went straight on to use another, keeping the length of the stroke long, thus ensuring the continuation of the deepening process. Short strokes would tend to remove glass less liberally from the centre of the mirror surface and there would be a tendency to preserve the long value of R, the radius of curvature, unnecessarily.

When the abrasive powder is placed on the upper surface of the grinding tool, it is usually put there as a paste made up with water. The grinding action can be judged by ear—there is a satisfying gritty biting noise accompanied by mechanical friction which has to be experienced to be appreciated. As the work proceeds, the gritty sound is softened and there is a subtle change in the friction between the two glass surfaces. This is a sure sign that the 'wet', as the working of one application of grit is called, is blunting and, maybe, drying up. Although there may be differences of opinion, a 'wet' usually lasts about 5 minutes. Because it is impossible to detect scratches and pits left behind by the larger grits until long afterwards, many workers operate 'by the clock' and most agree that six 5-minute wets of No 220, if properly applied, will remove any remaining No 80 (and No 120) pits and scratches. Searching the mirror surface with a magnifying lens is generally considered to be a waste of time in the early stages of grinding.

When the spherometer readings show us that the value of d is the value that we need to give us the all-important dimension, R, we may sluice down again and clear the decks for the finer stages of grinding. Six 5-minute wets of Nos F320, F400, and F600, ought to be sufficient to complete the fine grinding. Again, Ellison used No 500 in between the last two grades, but many regard this as unnecessary. Incidentally, the designation 'F' means that these powders conform to descriptions agreed by the Federation of European Producers of Abrasive Products (FEPA). While again some workers advocate even finer grits to follow the F600, I think that most are prepared to commence the polishing, having reached this stage. After all, the average size of the F600 particle is 9 microns at the beginning of the wet, so it will be very much smaller after 5 minutes' abrasive action. Indeed therein lurks a danger, and one that may entrap workers who go on to use even finer grades of powders. There is a risk of a 'seize-up' when the two glass disks suddenly stick together and refuse to part again. A gentle, lateral tap with a mallet, cushioned by a piece of wood on the side of the upper disk, will separate them, and then it is important that the mirror surface should be scrutinised carefully to ensure that no damage in the form of scratches has occurred. Should there be scratches, then, unhappily, there is no way out but to go back to one of the larger grits and work them out. Seizing can be avoided by shortening the duration of the wets or by replacing water for something else. Some workers mix some of their finer grades with alcohol, and add a few drops of water when the 'paste' is placed on the tool. The experienced worker can usually tell when seizing is likely to take place, and if there is any tendency, which sometimes manifests itself by a lack of smoothness of action, the work should be stopped immediately and more lubricant should be used.

PREPARING FOR POLISHING

We do not know at what stage our finely ground mirror surface is suitable for polishing. We have to base our assessment more on what we have done to it than on its appearance. The appearance, as has been pointed out, should be a uniform finely ground grey, and the magnifying lens should have shown it free of deep scratches. There is another check that can be made, but it must be pointed out that this is hardly a test because any finely ground glass surface should perform identically. Just the same it is an interesting and simple experiment to perform, and, to do it, we need an unfrosted electric light bulb—one in which we can see the filament. Having switched the bulb on, all we have to do is to hold the mirror with the ground surface uppermost, and, by oblique reflection, observe the image of the bulb filament. The image should be reddish in colour because only the red rays are being reflected specularly—the shorter blue rays are being scattered. If, by holding the mirror a foot or two away, the image of the filament is not seen, then it could be that the fine grinding is incomplete or that a larger grit has been used by accident. A post-mortem should be conducted without delay before hours of abortive polishing drive the message home.

Before proceeding to the polishing stage, everything should be cleaned down very thoroughly: all abrasives should be safely put away with their containers firmly sealed. The mirror and tool and supports and adjacent working areas should be hosed down and old cloths and such-like discarded. This is time well spent, and the job should be done thoroughly. Only then can preparations be made for the polishing of the mirror.

MAKING AND POLISHING LAP

The traditional method of making a pitch polishing lap is to melt the pitch in an old but clean saucepan and pour the contents over the grinding tool. To prevent the molten pitch from flowing away over the edges, gummed paper tape is stuck around the tool to form a low cylindrical wall about 0.3in high. Immediately after the pouring, the viscosity of the pitch increases and the gummed paper tape can be removed—normally the gum will not stick to the pitch. Then, while the pitch surface is still semi-fluid, the mirror, coated liberally with a mixture of soap and glycerin and water, or with rouge powder and water, is pressed down and worked for several minutes on the pitch. At no time should the mirror remain still during this process, for the purpose of the exercise is to produce a pitch surface that is as perfect as the spheroid of the mirror. By the time the pitch has hardened sufficiently to retain its shape, much of the residual pitch will be overhanging the sides of the tool where it can do no harm. The finished lap should be about ⅛in thick.

In the last section I described why the pitch lap had to be modified or cut into 1in squares, and I hasten to add that one does not necessarily have to resort to the task of cutting the V-shaped grooves across the lap. Such an operation must disturb the perfect shape that we have sought to produce and it can only be restored again, after cutting, by the cold pressing of the mirror against the lap. An up-to-date refinement is to use a specially designed rubber mat that contains square apertures separated by ribs. Immediately after the pouring of the pitch, the mat is placed on top and the prepared mirror surface is pressed down on this. After the pitch has set, the mirror can be removed and the rubber mat can be pulled clear of the pitch; the desired grooves will be imprinted in the pitch. The manufacturers claim that this is an improvement over the old method, but there are bound to be places where the pitch will need to be cut away, so in both instances the lap should be submitted to the cold press technique to restore the perfection of form that is so necessary before successful polishing can be accomplished. It should be added that the pattern of squares should not be symmetrical—the centre of the lap should be in the corner of a square. This helps to avoid the formation of 'zones' during the polishing stage, which will be dealt with later.

Before leaving the subject of the polishing lap, some mention should be made of the materials from which the lap is formed. Most optical workers devise and defend their own methods. There are several different kinds of pitch including English, Russian, Swedish, etc, and it seems to be a name given to a wide variety of viscous tarry liquids obtained from wood, coal and bitumen. So-called optical pitch contains other ingredients, such as rosin and beeswax, and, indeed, the easiest way is to purchase the material directly from the suppliers. A glance through the advertisements in Sky and Telescope and other astronomical publications soon reveals the names of companies that market this type of merchandise. Tempered Burgundy pitch for making polishing laps may be obtained from Edmund Scientific Co,

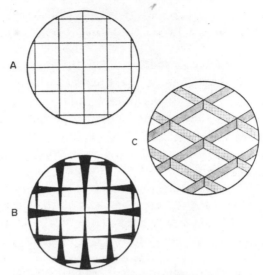

Fig 12 *How the pitch lap is divided into squares before the process of polishing commences. Notice that the centre of the lap is not in the centre of a square. B indicates how the squares are trimmed for parabolising. A close-up of pitch polishing facets is shown in C*

New Jersey. The more adventurous could try this formula for the manufacture of a suitable 'pitch':

Coal tar (melting-point 170° to 180° F)	2lb
Pine tar	4 liq oz
Beeswax	1½oz
Venice turpentine	2cc

Polishing pitch should flow slightly at room temperature, and should trim easily with a sharp knife. A long thin stick of pitch can be made by tipping molten pitch from a ladle on to a cold surface—it should be allowed to cool completely—then if the stick is bent slowly in the fingers it should bend without breaking, but if bent quickly it *should* break. If the pitch is judged to be too hard, then it must be melted again and a few drops of turpentine should be added and mixed in. Pitch that is too soft can be left to simmer for a while: in this way some of the more volatile substances will be boiled off and the pitch will be harder upon setting.

When it comes to the polishing of glass on pitch, there are certain danger signs to watch out for. For example, a lap that appears hard and glassy may be too heavily charged with rouge powder. It may also be found that such a lap does not polish as quickly as it should, and 'sleeks' may appear on the mirror surface. Sleeks are akin to scratches: they are shallow groove-like marks, and beeswax-coated tools are particularly prone to give this sort of trouble. Perhaps it should be mentioned that some workers make their pitch laps and give them a surface of bees-wax—such laps are not recommended for the beginner. Despite this, almost any polishing lap can produce sleeks, and one theory is that small balls of rouge and pitch break away from the lap and create these unsightly blemishes. Sleeks are best avoided by working the lap until the 'wet' is almost dry, a process which warms the lap and helps to regenerate the surface. It is obvious that glass polishing is an art, and it is certainly a subject where preaching should make way for practice, so let us press on.

THE FOUCAULT KNIFE-EDGE TEST

Having made the pitch lap we could commence polishing, but we cannot progress very far without devising some method of enabling us to measure the shape of the optical surface. So precise is the shape we are making that our measurements are no longer in thousandths of an inch but in millionths. Fortunately the apparatus that is designed to make these measurements is so simple, so cheap and easy to construct that no one need worry about it. The Foucault knife-edge test apparatus is almost described by its name, for a knife-edge, supported on a solid mounting, cuts the beam of light which is reflected by the vertically mounted mirror back to a focal plane close to the original pinhole light source. In practice, the light source is a pinhole with a frosted glass electric light bulb behind it: more elaborate light sources using condenser lenses may be used if required. The illuminated pinhole is set up in such a way that its distance from the mirror surface is equal to the radius of curvature, R. If the mirror surface is close to a spheroid, the reflected light from the pinhole will come to focus close to the source, and it will make an image the same size as the original pinhole. Indeed, under perfect conditions, the image of the pinhole would be reflected back on top of the original pinhole. Earlier I said that the radius of curvature, R, of a telescope mirror is twice the focal length and it may be confusing to read that a focused image of the pinhole is formed at a distance, R, from the reflecting surface. Allow me to qualify this by saying that this test, conducted at the centre of curvature of the spheroidal surface, is a special case. By placing our pinhole at infinity—in other words by using a real star—the distance of the focal plane is $R/2$.

This brings us to the crux of the whole business. Let us start with the image falling on the pinhole, as just described, and then we move the light source slightly to one side; the returning image will also move by the same amount, but in the opposite direction. If the eye is allowed to intercept the returning image, the mirror will be seen, and it will appear as a circular area filled with light, not unlike the full moon. So far so good. Now let us suppose that our mirror surface *is* a perfect spheroid: this would mean that the image of the pinhole would be assembled from reflected rays of identical length, that is, R. This is another way of describing a cone-shaped bundle of rays that converge to a minimum at the image plane, and cross over before they diverge ultimately to be lost in the distance. Placing our eye behind the image plane close enough to accept the marginal rays, before they diverge too far, we shall still see the mirror completely filled with light. If we go back too far, the diameter of the diverging cone of rays will be larger than the diameter of the pupil of the eye, and only selected parts of the mirror will be seen, depending on the position of the observer's head. With a 6in f/8 mirror and a pupil diameter of 0·25in, the maximum distance behind the cross-over point from which we would still be able to observe all of the mirror would have to be no more than 4in. If we position our eye an inch or two beyond the focal plane, we can bring the knife-edge into operation by moving it slowly across the bundle of rays until they are cut by it. At this stage we can partially abandon the written word and study Fig 13 and its caption. There are three possible ways that the knife-

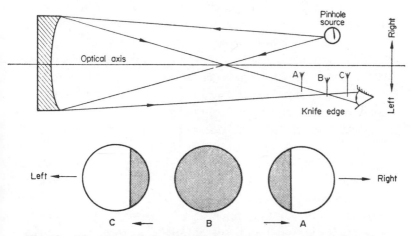

Fig 13 *The Foucault knife-edge test. Rays from the pinhole are reflected to an off-axis focus at B. Only marginal rays are indicated in the diagram. If the knife-edge at position A is moved to cut the rays, from right to left as shown by its arrow, the eye perceives the shadow advancing from the opposite direction as shown in circle A. At C the shadow moves across in the same direction as the knife-edge is moved. At B the knife-edge cuts the focal plane or cross-over point. If the mirror is a perfect spheroid, it will not be possible to determine the direction of the shadow movement: the mirror, as indicated in the circle, B, will darken down uniformly to extinction as the rays are cut off from the eye*

edge can cut the cone of rays. If the cone is cut well inside focus, that is, before the rays form the image, the eye will observe the black edge of the knife-edge advancing across the line of sight in a direction opposite to the direction that the observer is pushing the knife-edge. More specifically, if the knife-edge is being pushed across from *right* to *left*, the eye will observe the vertical shadow advancing from *left* to *right*. When the knife-edge is moved to a position behind the cross-over point, and closer to the eye, the observer sees the vertical black shadow moving in the *same* direction as he is moving the knife-edge. The important point is that either side of focus the direction of the advancing shadow is very easily determined. When the knife-edge cuts the cone of rays at the cross-over point it should be impossible to say from which direction the shadow moves. If the mirror is a perfect spheroid, the full moon appearance will just darken down to nothing as the knife-edge cuts across the focal plane. To answer the question of what the mirror surface should look like during the knife-edge test when the mirror is not a perfect spheroid, it is first necessary to discuss the subject of 'zones'.

Imperfections in mirror surfaces usually manifest themselves in the form of 'zones'. It is seldom that an isolated depression occurs in a specific spot on a mirror; indeed, it is difficult to see how such a thing could happen. If a depression is formed due to uneven polishing of the mirror, it will be a circular depression, and the circle will be concentric with the outer edge of the mirror. The mirror-maker calls this a depressed zone, and, if it turns out to be a deep one, it can cause some depression to the worker who has to remove it, for he must polish away the remainder of the mirror surface until it reaches the level of the zone. A raised zone is not nearly so troublesome to remove, for the mirror-maker can trim his polishing lap and plane down the offending zone until it is corrected. In this case the raised zone may be only a small fraction of the area of the whole mirror. In general, zones can be treated in a straightforward manner by scraping away the surface of the lap in those areas where the mirror is 'depressed' and by leaving active polishing lap surfaces where the mirror is 'raised'. Shallow scraping allows for subsequent cold pressing to restore the shape of the lap for further all-over polishing—in this way laps may be preserved and the chore of making new laps reduced.

The eye can observe raised and depressed zones very easily with the aid of the knife-edge. The flat, full moon appearance of the mirror will be disturbed by an annulus of light and shade which may give the illusion of being raised or depressed, or both; the effect is somewhat akin to the well-known optical illusions involving black-and-white squares which seem to make solid or hollow patterns. Also, a depressed zone does resemble a raised zone, for it is only in the interpretation of the shadows that we can determine which is which. In any case we have not discussed why we can see zones in this way. Tacitly we have assumed that the knife-edge has been playing its part, and that it has been cutting the true focal plane sufficiently to reduce the overall reflection from the mirror and to emphasise the zones. A zone is therefore an area of the mirror that reflects rays that do not share the same cross-over point as those coming from the rest of the mirror. A depressed zone reflects its bundle of rays to a focal plane in front of the normal cross-over point; a raised zone represents a reflecting surface belonging to a mirror of longer focal length, and its rays will find a focal plane behind the cross-over point. It is because the knife-edge is not at the cross-over of these wayward rays that the eye sees these zones with shadows on one side and highlights on the other. And the only way of recognising the difference between a raised and a depressed zone is to scrutinise the way the shadows are arranged. If a raised zone has dark shadows on the left and highlights on the right, a depression has the opposite—shadows on the right and highlights to the left. With practice and experience the mirror-maker will soon learn how to remedy the concentric raised and depressed zones that he perceives on his glass surfaces.

Apart from the zonal imperfections, a mirror can exhibit several different shapes on its way to becoming a paraboloid. These shapes can be identified according to the shadow pheno-

mena that are observed when the knife-edge cuts across the
reflected bundle of rays close to the cross-over point or points.

We have already discussed the appearance of the spheroid
under the knife-edge test, and are familiar with the flat, full
moon effect that gradually darkens out as the knife-edge is
advanced through the cross-over point or focal plane. A
spheroid is the only shape that can produce this all-over
darkening-down appearance. Other regular and common shapes
can be short-listed:

1. The oblate spheroid. The radii of curvature of the respective
   central mirror regions are longer than those in the peripheral
   areas. The shadow patterns formed during the knife-edge
   test could be mistaken for the paraboloidal shadows, the
   overriding differences being that the dark and light areas are
   reversed.
2. The ellipse or prolate spheroid. Here the radius of curvature
   is slightly shorter towards the centre of the mirror. Further
   deepening could convert this into a paraboloid.
3. The paraboloid. This, being the desired surface, is represented
   by a mirror form in which the radii of curvature progressively decrease in length towards the centre. With a small
   mirror of regular curvature, that is, without noticeable raised
   or depressed zones, it is necessary only to measure the value
   of the radius of curvature, R, for the peripheral zone, and
   compare it with the value, R, obtained for the cross-over rays
   reflected from the central region. When the mean values of
   many measurements of these two values of R are subtracted,
   and if the subscripts p and c refer to peripheral and central
   regions respectively, then:

$$R_p - R_c = \frac{r^2}{R}$$

The value of r in the numerator is equal to the physical radius
of the peripheral zone from the centre of the mirror. Going back
to our previous example of the 6in, 48in focus mirror, and

giving *r* the value of 3in (2¾in would be a better choice because this would represent the physical radius of the centre of the outer ½in wide zone), then:

$$R_p - R_c = \frac{3^2}{96} = 0.094\text{in}$$

The value 2¾in noted above would yield a smaller figure of 0.079in, so, if the difference between the two values of *R* lies between these two values, then all is well. Such a mirror is truly corrected.

The mirror-maker works on the premise that, if you cannot see a zonal region, you cannot measure it, and he makes from stiff cardboard, or any other handy material, a mask which isolates the central region and two diametrically opposite sections of the same peripheral region of his mirror. The finished product is easier to draw than to describe and such a mask is shown diagrammatically in Fig 14.

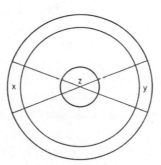

Fig 14   *A typical Foucault knife-edge testing mask. The diameter of the mask is the same as the mirror. In the case of a 6in, the central aperture,* Z, *may be about 1½in in diameter. Portions of the peripheral area,* X *and* Y, *are completely removed to reveal the reflecting surface*

4. The hyperboloid. To be avoided at all times; this shape occurs when the parabolising technique is carried to excess. The radii of curvature of the central regions of the mirror are

far too short, and we say that the mirror is over-corrected.
Such a figure is difficult to correct because, in a sense, it is
like having to treat a deep central hole, and hours of polishing on the outer regions of the mirror may be required to
bring the shape back close to a spheroid, whence parabolising
may be recommenced.

So much for the Foucault knife-edge test and the standard
mirror shapes. From the practical standpoint the apparatus
needs to be set up on very solid supports—creaky floors and
rickety tables should be avoided and, if no solid floor is available on which to place equally solid furniture, the best thing to
do is to bolt rugged shelves to the brick or stone wall structure
of the building in which the mirror-making activity is confined.
The lantern that provides the all-essential pin-hole can be anything at all—a tin can with a frosted light bulb inside and a tiny
hole pierced with a small sewing-needle will suffice. The wattage
of the lamp should not be too great, bearing in mind the proxi-
mity of the worker's face to it during the testing periods. The
knife-edge itself is more likely to take the form of an ordinary
3-hole safety razor blade which can be mounted in a variety of
simple ways. A 1in thick oak base measuring about 6×6in
will provide the means to support vertically a small metal
bracket on which the razor blade may be fixed. Any handyman
can solve these problems in less time than it takes to describe
them.

Lastly, and before we turn our attention to the polishing
again, it is important to note that we must allow the mirror to
come back to thermal equilibrium before it is tested. If ordinary
plate glass is used for the mirror—most workers prefer low
expansion material these days—the shadows can appear to be
hyperboloidal when the mirror comes straight off the polishing
lap. It is interesting to watch the shadows changing as the
mirror returns to room temperature after a few minutes. During
that brief period it can change from an over-corrected to an
under-corrected mirror.

POLISHING

Let us assume that the lap has been made, the facets have been cut, and that the cold pressing has restored the pitch surface to what we hope is the complement of the perfect finely ground shape of the mirror surface. Before we commence we need one more thing: a book in which to record all of the polishing activities. In this we must note—we can do this while the mirror is cooling off—all of the correct and incorrect things that we do: in fact the latter are more important than the former, for, if we jot down what we have done incorrectly, we can take remedial action. Nowadays, the operator could use a tape-recorder and write up afterwards when his hands are clean: this avoids sticking the pages together with pitch and covering them with rouge.

Initial polishing of a finely ground surface may take several hours. To reduce this period, some workers prolong the fine grinding by using smaller grits, but this is a personal choice. The first spell of polishing can proceed for about thirty minutes: the stroke should be kept short, and as the rouge and water dries out, the supply should be replenished. There is a good reason for keeping the stroke short, for one of the most difficult mirror defects to correct is what is known as 'turned edge'. To correct it, maybe as much as 90% of the central mirror surface has to be planed down until the flatter peripheral annulus is brought back to form part of the spheroidal shape. Keeping the strokes short prevents the turned edge from forming in the first place. In fact, many workers use subdiameter or smaller diameter polishing tools to correct turned edge, and such tools leave the peripheral regions almost untouched.

After the first spell of polishing, the mirror surface should be cleaned down—cotton wool or soft paper tissue can be used for this—and a high power magnifying lens or eyepiece can be focused on the outer regions. The outer regions naturally polish the more slowly, since they represent the 'overhang' region of the mirror during a considerable part of each stroke. If pits left behind by the grinding are much in evidence, the polishing process much go on until they have been removed. But after the first polishing period, a general view of the mirror will show that the central regions possess a better polish than the outer regions. If all goes well, the polishing should proceed in spells for several hours until the outer regions appear to be as polished as the centre. An inspection of the outer regions with the high-power ocular should show no worth while pits. However, it is only correct to point out that a few pits on a finished mirror do no harm—in fact, they are not nearly so troublesome as the ever-present dust that falls on the surface when the mirror is in the telescope.

Having got to this stage with remarkable pertinacity—most workers would have succumbed to the overwhelming temptation of examining the mirror with the knife-edge long before this— the worker should set the mirror up vertically on its easel. The easel, preferably, should have some means of fine adjustment. I use levelling screws through the wooden base: oak can be drilled and tapped to take ¼in diameter screws for this purpose. The easel is no more than a wide wooden angle-bracket, and the vertical member, which may be about 1in thick, should be wider than the diameter of the mirror. Ordinary round-headed wood screws can be driven into this near the base, and these will support the mirror; the general idea can be gleaned from Fig 15. But again the ways and means of doing this are legion, and in essence there is no reason why the Foucault knife-edge test should be performed along the horizontal. Less conveniently it could be conducted through the gaps that occur between ascending staircases, and then the mirror could rest on the floor facing upwards and without the need of an easel. (I have never heard of anyone using this technique for a small mirror, neither am I advocating its use. Indeed there are several disadvantages: heavy objects may be dropped on the mirror, and, unless it is solidly walled in, someone may walk on it, and then there is

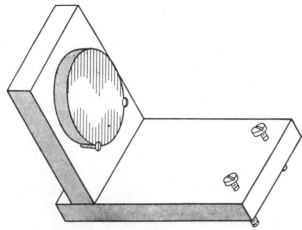

Fig 15 *A simple design for a mirror testing easel. It is made from hardwood and levelling screws may be used to make slight adjustments. A third screw or wooden 'heel' is out of sight behind the upright. The mirror is rested on two round-headed screws*

always the danger of the mirror-maker trapping his head between the bannister rails!)

Let us assume that these problems have been solved, and that the mirror surface has been submitted to the knife-edge test. If the shape approximates to a spheroid without raised or depressed zones, then with some rejoicing we could proceed to the stage of parabolising. Generally, the beginner would be most fortunate to see this happy state of affairs, and the shadow tests may show that his short polishing strokes have produced an oblate spheroid—by short we mean strokes that at the extreme position allow about one-third of the diameter of the mirror to overhang the tool. If this is so, then a further spell of polishing is required, and the stroke should be lengthened. Every few minutes the polishing should be halted and the mirror should be tested; it is fatally easy to remove too much glass and obtain the dreaded turned edge.

At this point, I would like to emphasise that our preoccupation with the polishing of the front of the mirror should not blind us to what is going on on the other side of it. I am alluding to the warmth of the hands of the operator which, transmitted to the glass, can change the shape of the blank sufficiently to upset the figuring of the optical surface. This is particularly troublesome with ordinary plate-glass mirror blanks. Low expansion glasses, naturally enough, are less prone to thermal troubles, and it is for these properties that they are selected in the first instance. Wooden knobs and disks, cemented to the back of the mirror with pitch to facilitate the handling during the grinding and polishing stages, sometimes lead to trouble because the heat transmitted through the glass cannot escape through the wood. Perhaps a better solution is to cover the whole of the back of the mirror with a disk of hardboard or plywood which can be cemented in place with pitch. This provides an all-over insulator for the heat trying to escape, and the heat that is trying to get in from the worker's hands.

Going back to the Foucault knife-edge test, how can we be sure that the interpretation of the shadows is correct? One simple way of doing this is to intentionally raise a 'bump' on the glass by expansion by gently warming it. In the case of ordinary plate glass the heat of a finger that has touched the glass for a few seconds is sufficient to produce a slight convexity which can be seen with the knife-edge test. On the low expansion glasses the heating would have to be a little more energetic—a small low-wattage electric bulb placed close to, but not in contact with, the mirror would suffice. If the set-up of the knife-edge test is such that the pin-hole source is off-centred a little to the right of the optical axis of the mirror, the reflected image will be formed the same distance to the left, where it may be examined. If most of the mirror surface is spherical and the knife-edge is adjusted to the true cross-over point of the returning rays, then the 'bump' will appear as a small area, consisting of dark shadow on its left edge gradually changing to highlight on its right edge. The dark and light areas show up against the overall 'half-intensity' background. This is how a bump will appear when the pin-hole source is to the *right* of the observer: if it is to the left the shadow tones will be reversed.

I am going to assume that we did observe a spheroid without any worthwhile zonal irregularities, and that we can press on with the parabolising. When we do this, we have to deepen the central regions leaving the outer areas virtually untouched; this, I may add, is easier to do than to remove glass around the peripheral areas and risk turned edge. And, in fact, this is one reason for not using the 'long-stroke method'. Possibly the best method for the beginner is to trim the facets of the lap so that the square areas are smaller around the edges of the tool but are about the same size in the centre. Fig 12 shows how this can be done, but remember that trimming is a process that spoils the surface of the lap and this means that cold pressing must be resorted to before parabolising can commence. The best way is to do the trimming and brush off any unwanted fragments of pitch or rouge. Place the mirror back on top with plenty of rouge and water in between, put three small and equidistant blocks of wood in position to prevent the mirror from sliding off the lap, and add a few pounds of weights to the mirror. Here we are talking about a small mirror where the addition of a few extra pounds would accelerate the process of cold pressing.

The four most used methods of parabolising a mirror are:

1. The long-stroke method.
2. Reshaping the pitch facets method.

The remaining two are:

3. The small polisher method.
4. Parabolising by overhang.

Of these, the small polisher method is used a great deal in the manufacture of large mirrors. We have already noted its usefulness in the eradication of turned edge, or in the prevention of turned edge in the first place. If this system is adopted for parabolising, the worker should make sure that his polishing spells are not too long. Failure to do this may mean risking the over-corrected figure of the hyperboloid, and the subsequent difficulties of working 'uphill'. The axiom must therefore be: 'little and often'—a little polishing punctuated at frequent intervals by knife-edge testing, remembering that is is far easier to work 'downhill' from a spheroid to a paraboloid.

Parabolising by overhang is capable of giving good results, and this method does not change the shape of the pitch lap. Also it is far more versatile. As the name suggests, the centre of the mirror blank no longer passes over the centre of the pitch lap during the stroke cycle. We can choose as much overhang as we require, and it hardly needs saying that the crescent-shaped area—depending on the position of the mirror and lap—which can be seen on the upturned pitch will be the same size as the invisible area on the under-side of the mirror. Thus as the three motions of polishing proceed, the central regions of the mirror will polish quicker than the peripheral areas. Again it is recommended that the operator should proceed cautiously, making frequent knife-edge checks and noting down all the time what he does and the results achieved.

ZONAL TESTING

As the parabolising proceeds, the Foucault knife-edge test must be used, not only qualitatively to assess the overall smoothness of curvature of the mirror surface, but quantitatively to measure the difference between the radii of curvature of the central and peripheral areas. A few pages back we worked out the difference between the values of R_p, R_c for a 6in f/8 mirror as:

$$R_p - R_c = \frac{r^2}{R}$$

and we conclude that if this difference is between 0·08 and 0·09in our mirror ought to be indistinguishable from a perfect paraboloid from the point of view of performance. In reality this means that when we scrutinise the mirror during the knife-edge testing we ought to make several—say six—attempts to obtain the exact radii of curvature of the central and peripheral areas.

(Often the word 'zone' is used instead of 'area' in this context—hence 'zonal testing'—but do not confuse this with the raised and depressed zones mentioned earlier.) Having got six measurements for both areas we can easily work out the mean values of R_p, R_c, and obtain the difference.

This is where the mask becomes essential, for the process of measuring the radius of curvature of the outer circle of the mirror means that we have to adjust the knife-edge position until the opposite outer edges of the mirror appear to darken uniformly. This process is difficult without the mask in front of the mirror to screen off the areas we do not wish to see, for the contrast differences, arising from intensity variations in adjacent areas, are sufficient to deceive the eye. The problem in the centre of the mirror is less acute.

From the practical standpoint, the difference between the two values of R amounts to this: the outer edge of the mirror is seen to darken evenly through the apertures in the mask. The position of the base of the knife-edge support is carefully noted—perhaps a line with a finely sharpened 'hard' pencil is drawn against the flat base and on to a sheet of white card which is glued or otherwise fixed to the shelf or table top that supports the pin-hole light source and the knife-edge apparatus. This line provides a measure of the value of R_p. So much for the peripheral region of the mirror. Now we can turn our attention to the central area which, being of shorter focal length, will require the knife-edge to be moved slightly closer to the mirror. The light source remains fixed, at least as far as this discussion is concerned. (Some workers prefer to mount the knife-edge and pin-hole on the same base and move the whole ensemble. Others employ 'beam-splitter' devices that reflect the returning light rays at right-angles into the eye: such a system remains 'on axis'.) After this brief digression we can return to the subject of the adjustment of the knife-edge until the 'null' or overall darkening position is found for the centre of the mirror. Again the hard, sharp pencil is drawn along the same edge of the knife-edge apparatus to provide the second line that is a measure of the value of R_c. Notice that we make no attempt to measure the true values of R_p, R_c and we do not have to worry unduly if these values are an inch or so longer or shorter than originally intended. The all-important dimension is the *difference* between R_p, and R_c, which is enshrined in the expression r^2/R.

So, recapitulating, providing that the curve of the mirror surface is virtually free of zonal imperfections—raised or depressed zones—and if the mean values of measurements of R_p and R_c lead to the required calculated difference then the task of parabolising can be considered to be finished, as are, nearly, these two sections.

Of course plenty of tasks remain to be done before the whole telescope materialises and all that I can do here is to list the main items:

1. *Newtonian flat*. This can be made or bought: and aluminised with the primary mirror.
2. *The telescope tube*. There is room here for plenty of individual solutions: cylindrical, hexagonal, square-section tubes are common. Even a 'no-tube-at-all'-type telescope can be made from a solid plank of hard wood with a right-angle bracket-type mounting, of similar material, for the mirror.
3. *Mountings*. The choice here is equatorial versus altazimuth and portable versus permanent. There is a great deal to be said in favour of a permanently mounted equatorial.
4. *Drives*. Manual versus motor (or clock)—for choice, a synchronous motor driven from a solid state variable oscillator with power output gets close to the ideal.
5. *Observatories*. Having an observatory means that all apparatus can be kept 'at the ready' and brought into action in the minimum of time.

5: TERENCE MOSELEY

Telescope Mountings

A TELESCOPE MOUNTING HAS FOUR BASIC REQUIREMENTS: IT MUST permit rotation around two axes which are mutually perpendicular; this motion must be perfectly smooth; the whole mount must be very steady; and the telescope must be able to view most, if not all, of the sky. To these may be added the practical aspects of ease of use, and ease of construction.

The rotation of the telescope around two mutually perpendicular axes enables it in theory to cover the whole sky in unbroken motion, but we shall see later that there are certain limitations to this. If one axis is vertical and the other horizontal, we have an *altazimuth* mount (from *alt*itude and *azimuth*), which is the simplest form, and may be preferred for some observations. If we tilt the vertical axis until it is parallel to the Earth's axis of rotation, it becomes a *polar axis* (p.a.) and the other axis will rotate around it in a plane parallel to the equator, giving us the *equatorial* mount, which has many advantages over the altazimuth, and is the one usually preferred. Since all celestial objects revolve in circles parallel to the equator, around the celestial poles, once the object to be observed is located in the field of view, rotation about the p.a. alone will keep it in view; this is not possible with the altazimuth, except at the north and south poles! In the equatorial, the other axis becomes a *declination axis* (d.a.), declination being the celestial equivalent of latitude.

THE ALTAZIMUTH
Apart from its use in specialised professional instruments such

as the vertical circle telescope and the Russian 236in reflector, the altazimuth is found mainly with the amateur observer, for two reasons. First, it is cheaper and easier to build than the equatorial, and so is often found on small 'first' telescopes. Secondly, some observers specialise in fields where the advantages of the equatorial are not required. These include artificial satellites, which usually move in orbits inclined considerably to the equator, and variable stars, where following the movement of the stars for long periods is not required.

Having used various telescopes on both types of mount for variable star (and other) observations, I must admit a strong personal preference for the equatorial, especially for very faint stars and when trying to find new ones. However, for those more gifted in its use than myself, and for the first-timer who doesn't want to attempt anything too ambitious, a brief description may be useful. The model described is a very simple one, as those who are capable of a more advanced version will be able to make the necessary adaptations to either the German or Fork types of equatorial described below.

This design is of wood, as those who require the stability of a properly machined metal mount will probably opt for the equatorial, or can adapt it themselves as already described. The vertical axis *A* (Fig 16a) is a piece of round dowel (e.g. stout

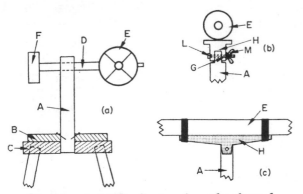

Fig 16   *A simple altazimuth made of wood*

shovel handle), about 2in diameter and 18in long. Two inches from one end is glued tightly the wooden disk *B*, 6–8in diameter and 1½in thick; this must be perpendicular to *A*. The bottom two inches of axis *A* pass through *B* into a close-fitting hole through disk *C*, 9in diameter by 2in thick. *A* should be as tight a fit in *C* as will allow it to rotate without gripping. The disk *B* takes the weight of the vertical axis and its load, and also provides stability in rotation which is controlled by *A* turning in *C*.

Disk *C* can be mounted on top of a tripod, on a pillar, or whatever is desired. The tripod has the advantage of portability, but for greater stability a heavy wooden post (4in square, or so) sunk about 2ft in the ground would be better. The height of the pillar or tripod will depend on the type (refractor or reflector) of telescope—5ft is about right for a refractor, 3ft for a Newtonian reflector.

Having got our rotation in azimuth, we now need a device on top of axis *A* to permit movement in altitude. There are several ways of doing this: a 1in diameter dowel *D* fits snugly into a hole about 2 in from the top of *A*, which carries the tube *E* at one end and a counterweight *F* on the other. Alternatively (Fig 16b), the top two inches of *A* can be cut down to give a flat-sided 'tongue' *G* about 1in thick. Over this fit two flanges, each ½in thick, of the tube saddle *H* (Fig 16c), which should be about half the length of the tube *E*. Through these flanges and tongue *G* is bored a ¼in hole to take a ¼in by 2½in Whitworth bolt *L* and wing-nut *M* (Fig 16b). Washers on either end of the bolt will prevent the wood splitting when the wing-nut is tightened to give the desired amount of friction. The tube can be held on the saddle by strong rubber bands if it is a light one, by leather straps, or whatever you find convenient. For a slightly larger tube, a vertical Fork-type mounting would be more suitable. All altazimuth mounts will be improved by the addition of manual slow-motions on each axis—these can be simple worm and wheel arrangements, operated by a rod on a universal joint, but they are not really worth while except for larger instruments.

THE EQUATORIAL

The origins of the equatorial can be traced back to a suggestion by Gruemberger in 1618, but it did not become common until its use by Fraunhofer in the early nineteenth century, in the form that has since become known as the German. There can now be distinguished seven basic forms of the equatorial mounting: as well as the German, we have the Yoke, English, Fork, Horseshoe, Split-ring and Springfield (including other similar types). They all have their advantages and disadvantages, and the choice will depend on such factors as type and size of telescope, financial and manufacturing resources, site, type of observation to be undertaken, and so on.

In the following section, I have not given construction details for each type, as this would involve a lot of repetition, and in any case, different individuals will probably have their own preferred methods of construction. However, I have given fairly detailed descriptions of two types, the German and the Fork, as a guide. These two are probably the most suitable for the amateur; those who wish to try other types can adapt and transfer the appropriate details according to the general principles outlined for each form. Obviously, details will differ according to the size of the instrument; in general I have taken as my model a 10in Newtonian of moderate focal length, and it should not be too difficult to make the appropriate conversions for a larger or smaller instrument.

German. The German type is perhaps the best known, and is particularly suited to refractors, from the amateur's 3in to the great 40in at Yerkes. It is in the shape of a T, in which the vertical stroke of the T is the p.a., set at an angle equal to the elevation of the pole. The cross-bar of the T is the declination axis, with the tube mounted at one end and a counterweight at the other.

The German mount is suited to all latitudes; it is relatively simple to construct, accessories are easily added, and the whole sky is observable. Against this may be weighed the necessity of a counterweight, which increases both the size and the weight of the whole instrument. In addition, when observing objects whose declination is greater than the altitude of the celestial pole, the telescope has to be reversed from one side of the pier to the other when the object crosses the meridian, both at upper and lower culminations. This is not too inconvenient for visual work, but it is impossible for photography, as not only would the exposure have to be interrupted, but the film or plate would have to be rotated through 180°.

A typical German design might be as follows. On top of a heavy concrete or metal pier, the p.a. *A* (Fig 17a) is supported at the correct angle by two bearings, *B* and *C*. The axis can be of mild or stainless steel, with the diameter depending on the load it has to carry—for a 10in reflector, 1½in would suffice, but the heavier this axis is the better, within reason. The bearings can

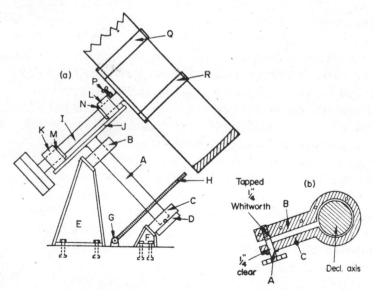

Fig 17 *A typical German mount*

be the conventional ball- or roller-bearing type, or they may be brass or phosphor-bronze, both of which have a very low coefficient of friction with steel. If conventional, it is advisable to get self-aligning bearings, unless facilities for their accurate alignment are available. Brass or bronze bearings will be cheaper, but again it may be difficult to line them up to the tolerance required for easy rotation. These bearings, which should be seated in aluminium housings, control the rotation of the shaft, but do not take much of the vertical thrust. To deal with this, a small hollow should be countersunk centrally in the base of the axis, in which rotates a steel ball of about $\frac{3}{8}$in diameter; this in turn rests on a steel or alloy plate D, which is bolted to the back of the bottom bearing housing. This ball can be located in position with a good dollop of thick grease in the centre of plate D.

The housings for the bearings should be bolted to the two supports E and F, which can be cast out of concrete (supports and base as one unit) with some reinforcing bars for strength, or they may be fabricated from $\frac{1}{2}$in welded steel plate as shown. For mounting these, see the appropriate part of the section on the Fork, p 94. Room should be left between the axis and the base plate for the worm G and wheel H to drive the telescope: even if this is not planned at the outset, you may well decide to do so later. The wormwheel should give as high a reduction as possible, and so it will be of large diameter—approximately the same diameter as the mirror is a good guide. So allow a space of half the aperture plus at least $1\frac{1}{2}$in for the worm and its mounting. It is possible to mount the drive unit on the axis above the top bearing, but this means extending the axis, and this introducing instability as well as raising the telescope height unnecessarily; it would also be more difficult to mount the worm in this position, and there is more likelihood of vibrations from the motor being transmitted to the tube.

On the top of axis A is welded a steel plate J measuring about 12in × 4in by $\frac{3}{4}$in; if this can be faced off on a lathe so that it is exactly perpendicular to *A*, all to the good. If not, shims can be placed under the bearing housings *K* and *L* so that the angle is correct. These housings, similar to those on the p.a., are bolted to plate *J*. Once again, the choice of bearings for the d.a., *I*, is optional, but self-aligning ones will be easier to install. The declination shaft should be about the same diameter as the p.a., and of such a length as to permit the tube to swing clear of the mount (including the wormwheel if fitted or planned) on the one hand, and to carry a counterweight at the other end. The lighter this counterweight, which can be of metal or concrete, the longer the shaft must be to balance the tube; the heavier it is the greater the weight on the p.a., but it can be mounted much closer to the p.a., giving more room and greater stability.

To prevent the declination shaft sliding up and down in its bearings, two collars *M* and *N*, diameter about $\frac{3}{4}$in greater than that of the shaft, and about $\frac{3}{4}$in long, to fit over the d.a., should be drilled and tapped 2 BA for two Allen set screws in each 180° apart. The collars should be locked in position on the d.a. by the set screws, either both inside or both outside the bearings, so that the shaft cannot slide up or down. If accessories are added to the tube, or its weight otherwise changed, balance can be restored by loosening the collars, moving the shaft slightly in the required direction, and re-tightening the screws. This can also be done by changing the position of the counterweight in similar manner.

To hold the tube steady in declination, some sort of locking device is required. The simplest way is to drill and tap for a 5/16in screw with a handle *P*, through the upper bearing housing; the screw bears on the declination shaft and stops it turning. If the bearings are of the ball or roller type, the housing should be extended by $\frac{1}{2}$in on the tube side, its i.d. the same as the axis diameter, and the hole for the screw drilled through this. Brass or bronze bearings can have the screw mounted directly in them. A better arrangement is shown in Fig 17b: when the screw *A* is tightened, the clamp grips all round the d.a., giving a much firmer lock. Side *B* is fixed to the bearing housing, *C* is free to move in and out. The clearance between the clamp and the d.a. should be 1–2 thousandths of an inch.

A declination slow-motion can be fitted to the d.a. if desired—a worm and wheel giving a reduction of about 200 to 1 will suffice. This can be driven by a rod and universal joint, flexible cable, or by a small motor for fine adjustments. The tube clamps *Q* and *R* should be fairly well spaced for stability, and if some provision is made for loosening and tightening them, the tube can be rotated to give the best eyepiece position (this obviously applies to round tubes only). Setting circles can be mounted on both axes if required.

Yoke. One of the disadvantages of the German is the necessity of a counterweight, which is not required by the Yoke (sometimes called the English, though this is in fact the version described below). The tube swings freely in the declination bearings *A* and *B* (Fig 18) on either side of a rectangular yoke forming the p.a., which is supported in bearings *C* and *D* at either end. No counterbalancing at all is required, there is no meridian reversal, and the design is extremely rigid. However, the polar region is permanently inaccessible, and so, to a certain extent at least, is

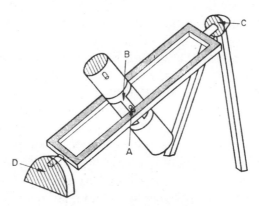

Fig 18   *A Yoke mount*

the area below it at any time. This is no problem for observers whose interest lies along the ecliptic, but another disadvantage is that this mount is unsuitable for refractors and Cassegrains, since the eyepiece would not only be very near the ground, but also under the yoke in most observing positions. Accordingly, the Yoke is most suited to large Newtonian reflectors, the best known example being the 100in Hooker reflector at Mt Wilson; but it is not to be despised for the small reflector, due to its ease of construction.

If the yoke is long enough for the tube to swing right through it, areas fairly close to the pole can be reached, if one does not mind a certain amount of obstruction of light by the upper bearing for the p.a. In addition, if the pier support for the upper bearing is replaced by a bipod as shown, the sub-polar area is not so obstructed as to prevent observation. This is particularly important for variable star observers, who may want to follow a star as continuously as possible.

A thrust bearing at the lower end of the p.a. takes the weight, as described for the German mount, though this one will have to be rather heavier, and can be installed in the short south pier, best and most conveniently made of concrete. The driving mechanism should be positioned between this pier and the bottom of the yoke, for convenience and unobtrusiveness. The other bearings for the p.a. may be chosen according to the principles outlined for the German. The yoke must obviously be rigid—for small instruments, say up to 8in reflector, it can be made of wood not less than 3×3in, well jointed, and treated with preservative. This will not do for large instruments, and in any case wood tends to warp, no matter how well preserved, sealed and housed, so a better alternative is a metal yoke. Depending on size, 3–5in galvanised pipe, with right-angle joints for the corners, might be chosen for availability, though one might experience some difficulty in fitting the declination bearings to the round pipe. Alternatively, steel girder of square, rectangular or *H* section, and sufficient thickness (at least ¼in), would be suitable.

The pieces are best welded together, but if this is impossible they can be joined by triangular plates bolted on at the corners.

As there are two declination shafts to take the weight, they can be of slightly smaller diameter than the single axis of the German. Again, self-aligning bearings will be easier to install, and provision should be made between the yoke and the tube for a slow-motion, declination clamp, and setting circles if required—the same general principles apply as for the previous mount. To stop the declination shafts sliding in their bearings when the yoke is turned to one side, collars can be fitted as described for the d.a., or thrust bearings on either side, as at the base of the p.a., as outlined for the German.

The north bearing support must be very rigid, and if a simple pier is chosen it must be sturdy—concrete with plenty of reinforcing (e.g. heavy wire, an old bed-frame) on a broad base is hard to beat. If the soil is soft, a suitable foundation must be provided (a local builder will give advice on this), or the pier will sink under its own weight. If the bipod type is chosen, heavy pipe sunk in the ground or in concrete foundations, and filled with concrete will be found satisfactory; as an approximate guide, the pipe should be about one-third mirror diameter, compromising between steadiness and amount of obstruction. If the latter is not an important factor it might be anything up to half mirror diameter.

Another disadvantage of this mount is its size, especially if it is to be housed in an observatory of some sort, and as the mount is definitely not portable (except for very small instruments, when the whole thing can be mounted on a wheeled base), some sort of covering is essential. If the observatory is a dome, the increase in size compared with a German or Fork is not great, but a run-off shed, slide-off roof or hinged roof type will need to be considerably larger. In view of the amount of obstruction afforded by the yoke, it is particularly useful to have a rotatable tube so that the eyepiece can be brought to a convenient position.

English. If the polar area must be clear, one solution is the English mount, also known as the cross-axis. Here the yoke is replaced by a single stout beam, with the telescope on one end of the d.a. and a counterweight on the other (Fig 19). The mount is quite rigid, as the p.a. is supported at both ends, and the d.a. only needs to be long enough on the tube side for the instrument to clear the upper bearing when pointing at the pole. Thus the pole is accessible, but not the area below it, subject to the qualification described for the Yoke. Since there is both a telescope and a counterweight, the p.a. needs to be very rigid, and this in turn requires a massive north pier. Examples of this type include the 72in reflector of the Dominion Observatory, British Columbia, the 82in McDonald reflector, and the 20in Double Astrograph of Lick Observatory.

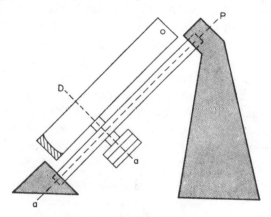

Fig 19   *The English mount*

Details of construction are much the same as for the Yoke, except that we have instead a single stout beam for the p.a.—pipe of about half mirror diameter would suffice, or square section hollow steel girder. If facilities are available, it could be made of a steel framework, like the jib of a crane. The d.a. will need to be sturdy, and once again a rotatable tube is desirable. This mounting is even larger than the last, as we have a counterweight to contend with, and so the same problems arise with housing it. Both these mounts, in fact, are more suited to low latitudes (where the north pier is not so high) than for those of Britain.

Fork. This is probably the most suitable mount for reflectors. Perhaps I am prejudiced, as my own 14·6in Newtonian is on a Fork, but if I give its pros and cons, the reader will be able to judge for himself. Firstly, the whole sky is always accessible, there is no meridian reversal, no counterbalancing is required, the mount is compact, and it is suitable for all latitudes. Against this may be set the fact that the weight of the tube and fork bear outside the p.a. supports, so there is a tendency to flexure. The Fork is not suited for refractors; Cassegrains can be used, especially if of the type with a non-perforated primary, where the light is deflected outside the tube near the base by a diagonal, but for observing near the pole with a conventional type a star diagonal must be used, and it is not very convenient.

Most amateurs have Newtonians, however, and if the p.a. and fork are made sufficiently rigid, the advantages of this type far outweigh those of any others. The 48in Schmidt at Palomar, the 60in Mt Wilson reflector, the 98in Isaac Newton, and the 120in Lick reflector are examples of the Fork mount. If the telescope balances near the mirror end, as most do, the fork arms can be fairly short, thus improving stability. In fact, if one is designing a telescope which is to be Fork-mounted it is well worth bearing this point in mind.

Since the Fork arms carry the telescope at some height above the base, it is usually unnecessary to mount a Fork on a pier—a heavy concrete block *A* (Fig 20) carrying the p.a. bearings can be bolted to a concrete base *B*, or they can be cast as one unit. The type and mounting of the p.a. bearings is similar to that described for the German, but the axis should be slightly heavier (1¾in for our 10in reflector), and the bearings spaced slightly further apart —say 14in in this case. It is a good idea to cast the top of the base at an angle of 20° or 25° to the horizontal so that the north bearing support *C* does not have to be too tall, and therefore unsteady. Room should be left between the p.a. and the base for mounting the wormwheel *D*, as for the German, which is why the base angle should not be equal to the latitude (unless no drive is to be used, or it is to be put elsewhere).

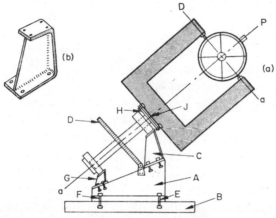

Fig 20 *A Fork mounted on a heavy concrete block*

It is easiest to cast the base block upside down in a wooden form built to measure, which is why I suggest having the block and its foundation separate. If the block is cast with two re-inforced flanges, each 2in wide × 2in deep, running along the E–W sides at the base, it can be bolted to the foundation with four heavy (about 8mm) Rawlbolts at the corners, *E*, *F*, etc. The holes in the foundation can be made with extractable mandrels of the right size when it is laid, or drilled later (this needs a powerful drill and large masonry bit); those in the flanges of the block are best formed when it is being cast, making them slightly larger than the diameter of the threaded part of the bolt to allow for adjustment in azimuth, although this can also be done by moving the bearing supports on the base block—in fact it may be safer to have both means of adjustment.

The bearing supports, *C* and *G*, can be cast with the block as a unit, with plenty of reinforcing, or they may be heavy steel plate. If the latter, they can be affixed to the base block by Rawlbolts, or just as easy and cheaper, by imbedding ordinary ⅜in × 3½in Whitworth bolts head downwards in the block when it is cast, leaving enough shaft projecting to bolt on the support, a possible design for which is shown in Fig 20b. Shims can be placed between these supports and the base, and/or between the bearing housings and the supports, for adjustment in altitude.

On top of the p.a. is welded a piece of ½in steel plate (*H*) 5–6in square—enough to clear either side of the fork base. A similar plate (*J*) is welded to the middle of the base of the fork, and the two joined by ½in Whitworth bolts at the four corners. Having the Fork and p.a. separate makes assembly and adjustment much easier. If it is not possible to face these plates off so that the fork is in line with the p.a., more shims will be necessary. Once aligned, mark two matching sides so that they can always be reassembled the same way.

The design of the fork itself is optional: in theory it is best to have the cross section increasing from the tips of the fork to the point of attachment to the p.a., but this is not easy unless you can make the whole thing yourself, or are prepared to pay for it being made to order. Methods of producing such a tapered fork include casting aluminium or brass, which can be done oneself, or iron, which needs a foundry, to the required shape, but expert advice, beyond the scope of this chapter, will probably be essential before this is attempted. It could also be made from steel plate welded to shape if facilities are available.

However, as long as the fork is sturdy enough, the taper is not really essential for amateur purposes, and it is much easier to make it in one of the following ways. As described for the yoke, 3 or 4in pipe is easily obtained and assembled. My own is made of welded ¼ × 4 × 4in square section steel tubing, which was chosen because the flat, mutually perpendicular surfaces are easy to work with. The declination bearings are probably best of the self-aligning variety, and the same principles apply as for the Yoke. Again, remember to leave room between tube and fork for a slow motion, declination clamp and setting circle, if required. Thrust bearings should also be fitted to the ends of the declination shafts.

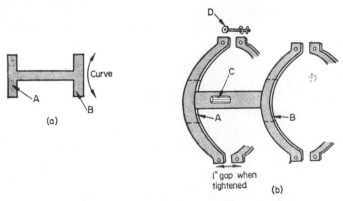

Fig 21 *Cradle design for a rotatable tube*

A rotatable tube is worth the extra trouble, and I will describe one way of doing this which I have found very satisfactory. I had a boilerworks roll two pieces of ¼in steel plate shaped as in Fig 21a, to a curve matching my tube outside diameter. Using the tube (which is itself of fibreglass for lightness) as a mould, I built up layers of fibreglass in strips, under and over the metal wings *A* and *B*, to produce a pair of tube clamps as in Fig 21b. The declination axes are then welded on (*C*, only one shown), and the two clamps joined by four ¼in × 2½in Whitworth bolts and wing nuts with washers, *D*, etc. Thus the tube can be slid entirely out of the clamps, or adjusted longitudinally for balance, and radially for eyepiece position, simply by adjusting the four wing nuts.

Horseshoe. A mounting which successfully combines most of the advantages of the Fork and the Yoke is the Horseshoe, designed by Russell Porter for the Palomar 200-in reflector. It is really a huge yoke, with the upper bearing enormously expanded, and cut away at the top, forming the 'horseshoe', so that the tube can nestle down into it to look at the pole (Fig 22). The advantages are: extreme rigidity due to the centre of gravity falling inside the bearings; no meridian reversal; and the pole is accessible. The disadvantages are: area below pole not visible and, especially for the amateur, difficulty and expense of construction. It is not suitable for refractors or Cassegrains (except very large ones—the 200in is often used as a Cassegrain), for the same reasons as the Yoke, and it also needs a large cover.

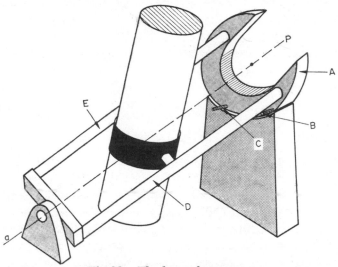

Fig 22   *The horseshoe mount*

However, for Newtonians it is probably the best mounting that can be built, and some amateurs have constructed their own. The design is essentially similar to that of the Yoke, except for the upper bearing, which consists of a large cylinder *A*, diameter about twice that of the mirror. This rests and rotates on two rollers, *B* and *C*, so that it can turn through at least 180°. A portion of this cylinder, which is centred on the pole, is cut away so that the tube can swing down into the bearing to look at the pole. The yoke-type arms D and E are attached to the cylinder, now the horseshoe, equidistant from the centre, on a diameter, forming the p.a. The tube swings in this yoke in the usual declination bearings. The distance between the rollers should be about half the cylinder diameter, giving maximum stability while allowing rotation through 180° E–W.

The chief difficulty with this mount is, of course, obtaining the horseshoe. Possible sources include discarded flywheels or old railway carriage wheels, but it is not easy to get one with an undamaged rim. If only one part of the rim is unusable, it can be the part cut out, but the remainder must be a very accurate section of a circle if the mount is to be worth the trouble. If the yoke arms are welded on, there is also the danger of the rim going out of true in the process. The north pier must be very massive, due to the size of the horseshoe, and like the Yoke and English types, this mount is more suited to lower latitudes, while a rotatable tube is again desirable because of the size of the mount.

Split-ring. This equatorial, designed by Porter, is, as far as I know, the only conventional mount that is entirely confined to amateur telescopes. This is because it is best suited to the smallish reflector (it is not suited for refractors or Cassegrains) —say up to 12in aperture—as for larger instruments it would be both very cumbersome and difficult to make. The Split-ring can be regarded as an evolution of the Fork, in which the upper p.a. bearing is greatly expanded into a hollow split-ring, *à la* Horseshoe, with the telescope mounted in its declination bearings inside it (Fig 23). The p.a. itself is replaced by a half-ring or U-shaped frame B, supporting the arms of the split-ring A, and rotating on the equivalent of the lower bearing of the old p.a., C. The advantages include: whole sky accessible, no meridian reversal, telescope weight inside bearing supports, and lowness. Disadvantages are: difficulty of construction and considerable lateral bulk.

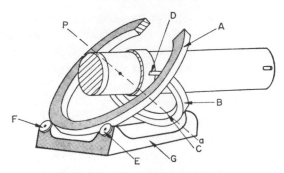

Fig 23   *The split-ring mount*

As with the Horseshoe, the chief difficulty lies in obtaining the split-ring, which needs to be rather larger than the horseshoe, aperture for aperture, as the whole instrument below the declination bearings has to swing through it, and there also has to be room for the declination shafts, *D*. In view of its large diameter it must be very massive to maintain the required rigidity, and obtaining, not to mention machining, such a large ring will not be easy. However, if something suitable is to hand, and the facilities for working it are available, the mount will prove very satisfactory in terms of rigidity.

The split-ring rotates on the two wheels or rollers, *E* and *F*, which need to be both accurate and sturdy. Bearing *C* incorporates thrust as well as rotation, and it is best for the three bearings to be mounted on one solid triangular concrete base, *G*. The tube, shown here pointing south, should be able to rotate in its saddle to allow access to the eyepiece. As with the Fork, the closer the balance point of the tube to the mirror end, the better, as this means a smaller ring.

Springfield. There are many different designs for mountings which have the telescope eyepiece, and therefore the observer, in a fixed position, but probably the best known and most satisfactory is the Springfield, invented once again by Porter. The optical and mechanical arrangement is shown in Fig 24, with the light from the primary being diverted along the hollow d.a. *A*, by the usual diagonal in the tube. Instead of being met immediately by the eyepiece as in the Newtonian, however, the beam is directed along the p.a. by another diagonal, *B*, where it is then examined by the eyepiece *C*. In Porter's original design, upon which Fig 24 is based, the axes are reduced to studs attached to circular plates *E* and *F* for stability and compactness. The tube is balanced by a counterweight *G*, placed as much out of the observer's way as possible. If a drive is required, a worm *H* can be mounted at the edge of polar axis plate *F*, which can incorporate or be made to carry a wormwheel.

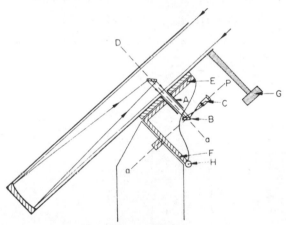

Fig 24   *The Springfield mount*

The mount is basically German in type, and therefore has the advantages of total sky accessibility and reasonable compactness, plus the added bonus in this type of a fixed observing position. There are several disadvantages, however: as well as meridian reversal, the image is given a mirror-reversal by the third reflection, and the latter also introduces a further light-loss. The extra length of the light path outside the tube requires a larger diagonal in the tube than with a straight Newtonian, which blocks light and causes image deterioration, and so

	Access to Pole	Access to Sub-Pole	No Meridian Reversal	Tube Inside DA Supports	DA Inside PA Supports	Compactness	Suitable for All Latitudes	Ease of Manufacture	Suitable for A, B, and C
German	✓	✓	×	×	×	✓	✓	✓	A, B, C
Yoke	×	× ?	✓	✓	✓	×	×	✓	A, C ?—
English	✓	× ?	✓	×	✓	×	×	✓	A, B, C
Fork	✓ C ?	✓	✓	✓	×	✓	✓	✓	A, C —
Horseshoe	✓ C ?	×	✓	✓	✓	×	×	× ?	A, C ?—
Split-ring	✓	✓	✓	✓	✓	✓ ?	✓ ?	× ?	A, ——
Springfield	✓	✓	×	×	×	✓	✓	× ?	A, ——

A = Newtonian; B = Refractor; C = Cassegrain.
'?' indicates that the quality concerned is not great, or applies only to a modified form.

short-focus instruments are unsuitable. Neither is it cheap or easy to make.

The characteristics of the various types of equatorials are summarised in the table. As was stated at the outset, the choice of a mount for a particular telescope is not automatic, but depends on many individual, as well as general, factors, but perhaps the information given here will make the decision slightly easier.

6 : H. K. ROBIN

Telescope Drives

ALL AMATEUR OBSERVATIONS OF THE SUN, PLANETS AND STARS ARE made from an Earth that is anything but stationary. The Earth rotates on its axis and it also orbits the Sun, while exhibiting a series of minor deviations from these general motions. Fortunately, only the main motions are large enough to need counter action by the amateur astronomer in the course of observations.

The Earth rotates on its axis once a day, so that, if we wish to make any celestial object appear to stand still in a telescope eyepiece, we must 'unwind' the Earth's motion by means of a telescope drive.

We have already seen that it is customary to mount a telescope equatorially, so that only one motion on the telescope is required to follow a celestial object. The true solar day is variable to the extent of 51 seconds between extremes, being 30 seconds over the mean solar day about 22 December and 20 seconds under about 17 September. A true sidereal day is the interval between successive transits of a star, and is the exact period of the Earth's rotation with reference to the star field.

In so far as it affects us, the mean solar day consists of 24 hours 3 minutes 56·555 seconds and the mean sidereal day consists of 23 hours 56 minutes 4·091 seconds. Thus, we must arrange to drive the polar axle at one revolution in 86,636·555 seconds for the solar rate and one revolution in 86,164·091 seconds for the sidereal rate. Both these figures are mean values, and are subject to further minor variations from several causes.

We are now getting into precise figures, so it is well at this point to consider what accuracy we really need in any practical system of drive. First, how 'still' do we want our image to remain? This depends on whether we are taking photographs or observing by eye. In practice we have to design for the extreme case, which is the taking of a long-exposure photograph. As has been explained in earlier chapters, the resolution of a telescope is controlled by the diameter of the object-glass or mirror; in other words, the aperture. Since we wish the drive to be as good as is required by the optical resolution, it follows that we must design the drive system so that the image is not noticeably degraded by movement in the time that it takes to get a good exposure.

Amateur telescopes vary greatly in aperture and photographic exposures may be from a few seconds to several hours. The table shows the precision of motion required for a wide range of circumstances.

Telescope Aperture in Inches	Limit of Resolution Arc-Secs	Accuracy of Drive Rate in Parts Per Million to Maintain Resolution for Exposures of:			
		1min	10min	1 hour	2 hours
1	5·45	60,000	6,000	1,000	500
2	2·73	30,000	3,000	500	250
4	1·36	15,000	1,500	250	125
8	0·68	7,500	750	125	62
12	0·45	5,000	500	84	42
15	0·36	4,000	400	66	33
20	0·27	3,000	300	50	25

Unfortunately, atmospheric refraction interferes with the image in two ways. First, the apparent position of a star is shifted from its true position in accordance with the zenithal distance. This is due to the varying mass of air through which we are observing. Secondly, local fluctuations of air density and temperature cause the image to wander in all directions about a mean position in the focal plane, so degrading the photographic image.

We now need to consider whether, in view of these serious drawbacks, we should try so hard to get a precise drive system. The answer must depend on what kind of work you want to carry out. Since an unsatisfactory drive is most frustrating, one should make a drive good enough under ideal circumstances to match the optical resolution.

Even when we have decided to make a drive of the required precision, we must still take account of the effects of atmospheric refraction. In all professional installations, means are provided for 'guiding' the image during a long exposure. This is done either by a second motion such as moving the camera on the telescope mounting or by small changes on the main R.A. driving rate in addition to slight adjustments on the declination setting. This can be done photo-electrically, as in professional observations, but, more often, it is done manually using great skill and patience.

The literature is rich in descriptions of ingenious methods of driving telescopes. At the turn of the century there were beautiful mechanical drives with heavy weights as the source of energy, and these gave way to electrical drives using mechanical governors of great precision. In turn, these were followed by synchronous motors running off the mains with complex gear trains to get both the solar and sidereal rates. The famous 100in at Mount Wilson was for many years driven by an oscillator made from a stretched wire similar to today's electric guitars. These old mechanical masterpieces now belong in museums, and have little relevance in today's astronomy.

While I appreciate that every amateur has his own preferences when it comes to mechanical versus electrical devices, I do believe that the only practical method is to use a synchronous electric motor driven directly from the mains or by an oscillator through a power amplifier. This is the method we shall now discuss.

Referring to the table, we see that the stabilities vary from six parts in one hundred for a short exposure on a small telescope to twenty-five parts per million for a two hour exposure on a 20in instrument.

The electric grid system has a good long-term accuracy of frequency in that the time given by mains clocks is within a few seconds over twenty-four hours, but the minute-to-minute rate may vary from -2% to $+2\%$, depending on the total load on the system. This means we could not use the mains for a telescope drive on, say, a 4in telescope for longer than a minute or so. Remembering that the solar and sidereal rates are about 3% different it follows that beyond a 3in telescope we must make separate arrangements if we want to make long exposures. However, it is fortunate that since the solar exposures are so very short, we can reasonably forget the solar rate and use only one drive at the sidereal rate.

Oscillators suitable for telescope drives can take various forms, such as electro-mechanical such as tuning-forks and stretched wires, resistance capacity oscillators, inductance/ capacity oscillators and quartz-crystal oscillators.

Stabilities vary over a wide range; a very good fork can give one part in 10^5 stability and a crude one without temperature correction will give one part in 10^3. A resistance/capacity oscillator carefully made with the right choice of materials will give one part in 10^3 and an inductance/capacity oscillator using a well-matched pair of LC* will approach five parts in 10^3. The quartz oscillator is very much more stable, and can reach one part in 10^8 when carefully ovened with a good temperature control and even without an oven it can give one part in 10^5.

For a fixed observatory carrying out work needing long exposures the quartz oscillator is undoubtedly the best choice. For those who wish to have an accurate drive at both rates, the problem of getting two rates from one crystal has been over-

* L = inductance (in Henrys); C = capacitance.

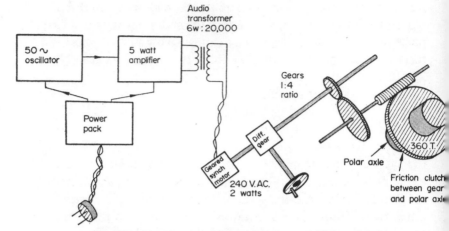

Fig 25 *Components required for an oscillator drive*

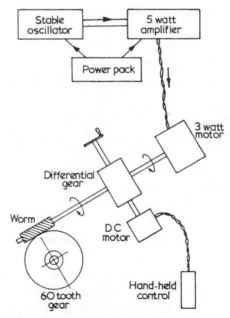

Fig 26 *Components for Oscillator Driven System with manual or motor control of differential motion*

come by an electronic dodge by which the faster rate is obtained from a solar rate by feeding back extra pips in the frequency divider chain at controlled rates so that the motor runs at the faster sidereal rate. This method is well within the capability of many Hi-Fi enthusiasts and when achieved is by far the most accurate system.

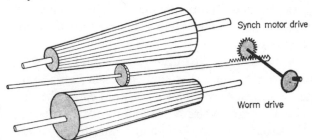

Fig 27   *Form of controlled ratio friction drive for use in a
fixed frequency system used at Mount Wilson*

There is now available both in the USA and the United Kingdom a wide choice of transistorised oscillator drives with control over the frequency covering the range of solar and sidereal rates for driving 2/5 watt small synchronous motors. These are also very useful for field work, where the power can be taken from a car battery. Small synchronous motors with integral gear boxes can be obtained from manufacturers and many 'surplus stores'. They have outputs at various rates, and one rev per minute, five revs per hour, and fifteen revs per hour are the most common.

We have discussed the requirements of the supply of power to the synchronous motor, but what about the reduction of the one rpm or so that we get from the motor to the one rev per day required by the polar axle? This is one of the most difficult problems facing the amateur telescope maker. Backlash and friction are the bugbears, problems that must be squared up to with the greatest determination.

In the 200in at Palomar, friction was virtually reduced to zero by floating the north end on oil pads. This produced its own problems in that vibrations caused by movements of the observer took 2 minutes to die down sufficiently to expose a plate. Artificial friction in the form of dampers had to be built into the structure. The motor drives the polar axle by means of a 20ft diameter worm gear. So we see that the design is aimed at removing friction and minimising backlash by using an enormous driving gear. The same problem exists even at the small sizes in the amateur range.

The trouble is that, with inevitable friction and with motion at such a low rate, the polar shaft tends to move in a series of small jumps in response to the force applied by the drive gears. Bearing in mind the accuracy we want (and that one arc/second equals one thousandth of an inch at 17ft radius), we see that the linear motion of the polar gear wheel is going to be very small indeed (about half a thou per second on a 12in diameter gear). This in turn means that the worm shaft driving the worm gear (which should be as large as you can afford) must have negligible end float. Practice dictates that the end float must be removed altogether by heavy spring loading applied through a thrust ball race.

The meshing of the worm and wormwheel should also be spring-loaded, calling for careful choice of materials and lubrication if the resulting friction is not to be excessive. All of this work should be done with the same care and devotion normally lavished on the optics if a drive matching the telescope resolution is to be achieved.

Means will have to be provided for declutching the drive when manually slewing the telescope. Here we find that ingenuity applies again. One way is to mount the wormwheel freely on the polar axle and clamp it against a flange when the drive is required. This should be done so as to avoid any slop between the gear central boss and the polar shaft, which would allow eccentricity leading both to backlash and an incorrect angular drive rate. Some firms supply complete wormdrives including a clutch; and for many amateurs without access to machine tools, this may be the only way to get a drive good enough for their requirements.

In the foregoing we have assumed that gears would be used for transferring the motor torque to the polar axle. In fact, friction drives are possible and in some ways preferable. At the IAU Symposium (27) on large telescopes, several workers in the field described loaded smooth wheel drives that have performed well. The three advantages of this method are that the high cost of precision tooth cutting is avoided, that the simple declutching is achieved by removal of the loading and any odd ratio can be obtained by control of the two diameters. I would certainly advise amateur telescope builders to give this simple method some thought.

We mentioned earlier that a second motion was generally used for the purpose of guiding. This second motion can also be used for moving the field of view about a mean position while the sidereal drive is running. There are several ways of doing this. In one good method, a differential gear box is interposed between the synchronous motor and the worm drive so that extra rotation on top of the sidereal drive can be added or subtracted by a hand-wheel or, more elaborately, by a DC variable speed motor. A companion motion in declination must be provided to give full freedom in all directions to the image. Another way, more attractive to electronically-minded people, is to control the frequency of the driving oscillator by large amounts; but since most synchronous motors will not easily run at very low or very high frequencies, the slewing is limited to the rather plus or minus 15 arcsecs per second of the sidereal rate. The former method is to be recommended, since it sets no limit to the slewing rate or direction.

This chapter would not be complete without a few words about setting circles and readout systems. How accurate should your readout be? Again it all depends on what work you want to do with your instrument. One thing is certain—it is not possible to readout from circles sufficiently accurately to get any particular object into your field of view except in the smallest telescope. Most scales read to 1 degree only so that at best this will only get the object into view in the finder. The R.A. and Dec are then manually adjusted to bring the image into view in the eyepiece of the telescope. This may be sufficient for most amateur work, but for those who desire something better, here are a few notes on what is being done by the professionals.

If the main wormgear is made with 360 teeth, then each revolution of the worm will give 1 degree and if this drive shaft has a scale of 60 divisions, each division will be 1 minute of arc. Another set of divisions must be chosen if you wish to readout in hour angle (15 degrees equals 1 hour). This will give hour angle information which must be added to the sidereal time to give true R.A. readings. This procedure can be carried still further on earlier gears, but backlash will soon dominate, making further accuracy impossible. But remember that if this sort of accuracy is required, the equatorial mount must itself be aligned with the Earth's axis to a similar precision, and this is not at all easy.

Modern big instruments use digital shaft encoders (optical devices) on the main and earlier shafts and these read remotely on a dial or digitally on a detached consol. These encoders are very costly for the most accurate types; however, much lower accuracy encoders are not impossible for the amateur to construct.

Even though we can not as amateurs reach the ultimate in precision, it is worth using gear systems based on 360:1 and 60:1 so that a differential reading in degrees and minutes can be obtained between a known star and an unknown one, making it possible to record an unknown star's R.A. and Dec. In my own instrument I have fitted synchros to the second shaft, and can read out on clock faces the difference in R.A. and Dec from known stars. If the clock faces are set to the figures of the known star and then the synchros switched on the dials will follow the telescope slewing to any point in the sky and readout to within 1 minute of arc.

7 : GILBERT E. SATTERTHWAITE

Adjustments to the Telescope

THE NECESSARY ADJUSTMENTS TO THE TYPES OF TELESCOPE USED by the amateur fall into two categories—those which must be carried out when the instrument is set up, and those which may have to be carried out from time to time as part of routine maintenance. Much of the precise adjustment of a telescope is a matter of delicate optical alignment requiring much knowledge and skill; in this chapter we are concerned only with those procedures which the amateur can himself undertake successfully, with little danger of damaging the instrument providing that he tackles the job with care and common sense. It is perhaps worth emphasising here, however, that any good telescope is a carefully designed piece of apparatus, containing delicate and expensive optical components in precisely calculated relative positions. Such components are easily damaged, and their positions can be disturbed by careless handling. The mounting too, though it be of sturdy design and finely engineered, can easily be damaged by rough treatment. It follows then that any telescope should be handled with great care at all times, and should never be dismantled or otherwise tampered with save by someone with knowledge and experience of the adjustment of astronomical equipment. The trouble—and possibly expense—of arranging for help and advice from such a person is a small price to pay in order to avoid the disappointment of an instrument ruined, or at least requiring expensive repairs.

The principal adjustments required when setting up a telescope are described, including the setting of an equatorial head and the alignment of a finder. The optical adjustments of the three most common types of amateurs' telescope are outlined, and finally some hints on routine maintenance are given.

SETTING UP A TELESCOPE

Portable instruments. Most portable telescopes are mounted on a folding tripod, usually of wooden construction; some may have a more rigid mount and should perhaps be classed as 'semi-portable'. Setting up instruments of this latter kind may require the help of a strong friend.

Few problems are normally encountered in setting up a portable instrument. If it has an altazimuth mounting, beyond ensuring that it is firmly set up on level terrain, there are no special requirements. Many portable and semi-portable telescopes have equatorial mountings, however, and as far as is practicable they must be correctly aligned and adjusted in the same way as a permanent instrument. The user should at least ensure that the polar axis is inclined to the horizontal by an angle approximately equal to his latitude, and that it is aligned north-south. A good pocket compass is helpful here, to determine magnetic north; to obtain the direction of true north the current magnetic variation must be applied.

Permanent instruments. A few telescopes with altazimuth mountings are permanently installed, but these require no special adjustments; the erection of a permanent equatorially mounted instrument is however a much more complex business. One of the most important adjustments must be considered when the instrument is first erected: this is the alignment of the polar axis in the meridian plane, that is, it must lie on a north-south line. The larger instruments usually have some provision for adjustment in the equatorial head itself, which enables the head to be rotated in azimuth and locked in the chosen position.

Sometimes this movement can be finely controlled by means of an adjusting screw. If the instrument to be erected has such provision, the head should be set in the middle of its run of permitted movement before erection commences.

Large telescopes usually require the construction of a concrete pier to carry the mounting: before this is made it is desirable to establish the north-south line across the proposed site. To a reasonable approximation, this can be done with the aid of a good magnetic compass. The line obtained will be the current magnetic north-south line, and the required meridian must be obtained by applying the current value for magnetic variation. In Britain, for example, this can be obtained from the latest edition of the 1in Ordnance Survey map of the district, where the value for a given year and the annual variation are given.

Once an approximate north-south line has been established, it is a good idea to mark it with a post either side of the telescope site and sufficiently far from it not to impede the work of erection. The line can later be determined with greater precision by observation of Polaris, the north pole star. The observation must be carried out at about the time of upper or lower transit of Polaris across the meridian. These times can be determined from the local sidereal time: upper transit of Polaris will occur when the LST is equal to its right ascension (about 2h 5m in 1972), and lower transit 12 hours later.

A plumb-line should be suspended at each end of the approximate line, and then by sighting across both plumb-lines simultaneously, one of them should be moved until they are exactly aligned on the star. Another way of checking the meridian line is by observing the direction of the shadow of a plumb-line at local noon (12h GMT, plus the appropriate corrections for longitude and for the equation of time on the date concerned).

Once the meridian has been established, the telescope mounting should be placed in position and adjusted very carefully so that the polar axis is aligned along the meridian before the base is fastened down.

When the stage of final adjustment is reached, the initial setting of the azimuth adjustment of the head in the middle of its run, and the careful alignment of the mounting, should ensure that there is sufficient movement available to permit the polar axis to be aligned in the meridian with great precision.

It is also very important that the polar axis is inclined to the true horizontal by an angle exactly equal to the latitude of the site. Many larger equatorial mounts are provided with some adjustment for this—either freedom for the plate carrying the polar axis bearings to be moved in the vertical plane and clamped, or provision to tilt the whole mounting. Smaller equatorial mounts often have adjustable feet, but many have no such provision, and if used in a latitude significantly different from that for which they were designed, have to be adjusted by means of suitable blocks or other packing beneath the feet or baseplate.

THE PRECISE ADJUSTMENT OF AN EQUATORIAL HEAD

To function properly an equatorial head must be adjusted so that the following criteria are met:

(1) The polar and declination axes must be exactly perpendicular to each other; this should be the case if the instrument has been properly manufactured, and no adjustment is normally provided. The optical axis must also be perpendicular to the declination axis, and this may require careful adjustment of the optical parts by a skilled person.

(2) The polar axis must be exactly parallel to the Earth's axis of rotation, that is, it must (a) lie in the plane of the meridian and (b) be inclined at an angle to the horizontal equal to the latitude.

(3) If setting circles are fitted the zero reading of the index marks or verniers must be set.

In most instruments (1) cannot readily be adjusted, but if the instrument has been properly cared for it should remain as set by the manufacturers. If the brackets carrying the bearings for each axis are on separate parts of the head before assembly, care must be taken when bolting them together to achieve equal pressure across the area of contact.

The coarse adjustment of (2a) takes place during erection of the instrument and has already been described; that for (2b) is carried out by visual inspection with the aid of a protractor or a template cut to the local latitude. A spirit-level or small plumb-line will also be required. The fine adjustments of the polar axis and the circles must be carried out in a specific sequence, as follows:

(i) zero setting of declination circle index
(ii) final setting of polar axis in altitude
(iii) final setting of polar axis in azimuth
(iv) zero setting of hour circle index

(i) Setting the declination circle. The fine setting of the declination circle index or vernier is done by observation of a fairly bright star; one should be chosen that is fairly close to the meridian at the time, so that its altitude will not change significantly during the period of the adjustment. The telescope should be positioned on the east side of its mounting,* the star brought to the centre of the field and the reading of the declination circle taken. The instrument should then be moved to the corresponding position west of the mounting, the star again brought to the centre of the field and the declination circle again read. The index mark or vernier should then be adjusted to indicate the mean of these two readings. The process should then be repeated and any small residual error again corrected. The declination circle should then give the same reading in both positions of the telescope. As a check the entire process should be repeated using a star of quite different declination.

---

* This applies only to asymmetrical mountings, e.g. German and English. For others, e.g. Fork and Yoke, rotate through 180° in R.A. and then in Dec, and take the record reading.

(ii) Altitude setting of the polar axis. The final adjustment of the polar axis in altitude can be carried out in several ways. If the instrument is equipped with circles, the simplest method is to choose a star of known declination and small zenith distance. Set the declination circle to the declination of the star and clamp. The elevation of the polar axis is then adjusted until the star can be centred in the field by sweeping in right ascension only. The operation is then repeated with the telescope on the other side of its mounting and any further slight adjustment made. As a check the whole operation should be repeated using a different star.

If the instrument has no circles, a quick setting can be made which makes use of the fact that the north celestial pole lies about 51' from Polaris, in the direction of η Ursæ Majoris, the star at the end of the handle of the Plough. At approximately six hours before the meridian transit of η UMa (R.A. 13h 46m), and again six hours after its transit, the altitudes of both η UMa and Polaris are approximately equal and are within one minute of arc of that of the pole. This method must therefore be used six hours before or after culmination of η UMa.

The declination axis must first be set in the vertical plane through the polar axis, using a plumb-line, and the polar axis then clamped. The elevation of the polar axis should then be adjusted until Polaris can be brought to the centre of the field by sweeping in declination only.

A more accurate method, which calls for some patience, requires an eyepiece with cross-wires which must be oriented north-south and east-west. This is done by setting the telescope to point due south, and allowing an equatorial star (altitude 30° to 45°) to trail across the field. The eyepiece is rotated in its draw-tube until the star trails along the horizontal wire; this wire is then oriented E–W.

Two stars must then be selected about six hours east and west of the meridian, and of declinations between + 40° and + 50°. The telescope should be set on the easternmost star, which is then followed by means of the clock drive or manual slow motion R.A. If the elevation of the polar axis is slightly incorrect, the star will drift slowly away from the horizontal wire: some time will be required before this effect becomes perceptible. The observation is then repeated on the westernmost star. The apparent drift of the stars indicates the correction required to the position of the polar axis as follows:

E star drifts north, ⎫ Elevation of polar axis should
W star drifts south ⎭   be reduced
E star drifts south, ⎫ Elevation of polar axis should
W star drifts north ⎭   be increased

The amount of adjustment required can only be determined by trial and error; this method gives considerable precision, however, and with patience enables this particular maladjustment to be eliminated.

(iii) Azimuth setting of the polar axis. If the telescope has circles, the polar axis, now correctly set in altitude, can be corrected in azimuth by the following procedure: choose a star of known R.A. and Dec at an altitude of 40 to 50° above the eastern horizon. Set the declination circle to the Dec of the star and clamp. The polar axis is then rotated in azimuth until the star can be centred in the field by sweeping in R.A. alone. This procedure should be repeated using a similarly placed star west of the meridian.

If the telescope has no setting circles, the azimuth of the polar axis can be adjusted by a procedure which must be carried out at the time of upper or lower transit of η Ursæ Majoris. The declination axis must be set horizontal, using a spirit-level, and the polar axis clamped. The azimuth of the equatorial head is then adjusted until Polaris can be centred in the field by using the slow motion in declination only. If the adjustment is made when η UMa is at upper culmination, the north end of the axis will be about 1′ west of the correct position; at lower transit it will be about 1′ too far east.

(iv) Setting the hour circle. The declination axis should now be set exactly horizontal using a spirit-level, and the polar axis clamped. The hour circle index or vernier should then be set to read zero. Many instruments have two verniers, a fixed one to read local sidereal time and a movable one to indicate the R.A. of the object observed; in such instruments both verniers should be set to read zero.

An alternative, and more accurate method, is to calculate the exact time of transit over the local meridian of a star of known R.A.; using a high-power eyepiece with cross-wires, the star is observed shortly before its transit. It should be set exactly in the centre of the field, using the cross-wires, both axes should then be clamped and the star kept in that position by use of the R.A. slow motion. At the calculated instant of meridian passage, the observer must stop the slow motion; the hour circle should then be set to read zero and the R.A. vernier, if any, to the R.A. of the star.

ADJUSTING THE FINDER

Most finder telescopes are fitted with cross-wires in the focal plane of the eyepiece; when the finder is properly adjusted any object upon which the finder cross-wires are set should appear central in the field of the main telescope. Even when a very high-power eyepiece is used, the object should at least be somewhere in its field. Clearly, therefore, the finder must be adjusted relative to the main telescope with considerable precision, and once adjusted should remain undisturbed by subsequent use of the instrument.

Many finders are fitted with a pair of 'wires' set perpendicular to each other, thus forming a simple cross. In using this type of finder the desired object has to be slightly offset, in one of the four 'corners' of the cross. A far better system is the double cross formed by two pairs of parallel wires; with this type the object is centred in the 'box' at the centre of the cross-wires. [The term 'wires' is used here, and indeed the cross is often formed of fine, stretched wires; it is sometimes formed by black lines photo-printed or etched on a glass plate, however, or even (as in the precision instruments used in positional astronomy) spider's webs.]

The adjustment of the finder cross-wires to coincide with the centre of the field of the main telescope must be provided in the finder mounting. This usually consists of a pair of ring brackets, but many cheaper instruments have a one-piece mounting for the finder. This should be provided with three adjusting screws towards one end and set at intervals of 120° so that the direction of the finder can be varied; one end of the mounting bracket usually acts as a fulcrum, but accurate adjustment of a finder mounted in this way may prove difficult.

Where two ring brackets are provided, either one or both of them will have three positioning screws. Where only one bracket has them, the other will have a smaller internal diameter and will be shaped to provide a snug fit for the finder tube and to act as a fulcrum for it. The best form of mounting for really accurate adjustment is a pair of ring brackets in which both are provided with adjusting screws. The screws should all have locking nuts.

The adjustment of the finder should be carried out from a site with a good distant view, using a suitable object in the far distance. A church spire or factory chimney are ideal. A low-power eyepiece should be used in the main telescope, which should be set on the chosen object by trial and error and clamped in position. A suitable part of the object, for example, the tip of the spire, should be set exactly in the centre of the field. The three adjusting screws in the finder-mounting bracket farthest from the eyepiece should be set so that they clamp the finder tube exactly in the middle of the ring. The finder is then moved, by loosening and tightening opposing screws in the nearer bracket, until the chosen part of the object is centred on the cross-wires. The final adjustment is often made easier if slight adjustments are made to the screws in the forward bracket also. When the finder is precisely set, all six screws should be checked to ensure that they grip the finder tube firmly: no movement should be possible, otherwise vibrations during use will soon disturb the setting. Finally the locking nut on each screw should be tightened.

A useful check on the final adjustment is to change to a higher power eyepiece in the main telescope: the object being viewed should still be close to the centre of the field.

As soon as possible the finder should be tried out at night, using a naked-eye star as the test object. Unless a really suitable distant object has been used for the initial setting, it will often be found necessary to make a further slight adjustment to the finder alignment. If so, this must be done with care and fairly quickly, because of the diurnal motion of the star; it is a good idea to use Polaris for the adjustment in order to minimise the need for haste.

ADJUSTING THE OPTICS

A telescope is basically an optical system—that is to say, a train of optical components. The tube, draw-tube(s), mounting, etc are merely means of maintaining the optical train in its correct arrangement and directing it towards the object to be viewed. Clearly, then, the performance of any telescope depends entirely on the correct positioning and alignment of the optical components; however well made and precisely adjusted its mechanical components may be, no telescope can perform satisfactorily unless its optical train is properly adjusted. In the main, correct adjustment of the optical components means that they are all centred on a common axis—the optical axis of the instrument—and 'squared on' to the axis so that a pencil of light passing through the centre of each component would travel, undeviated, along the optical axis. (It should be noted that the optical axis may itself be deviated, as in, for instance, a Newtonian reflector.) The adjustment of the optical components to achieve this state is termed *collimation* of the telescope.

Another important consideration is to ensure that the separation of the various elements along the optical axis is correct.

In any good telescope the mountings of the optical elements are designed so as to maintain them in correct alignment with a minimum of adjustment; it is however sometimes necessary to make certain adjustments—especially in the case of reflectors. The normal collimation adjustments of the three types of instrument most commonly used by amateur observers are described here.

The refractor. Most refractors are professionally made, and if of good quality need little adjustment. The only collimation adjustment normally required consists of squaring-on and centering the object glass. Imperfections in the image produced by an object glass that has been properly centred and squared-on are most likely due to incorrect positioning or alignment of the components of the object glass in their cell; should these components require adjustment it is best entrusted to an experienced optician. The existence of such maladjustment should be suspected if, for instance, the out-of-focus image of a star is elliptical rather than circular, or is otherwise deformed.

Most refractors of 4in aperture or less have no provision for adjustment of the O.G. cell; it is usually threaded for screwing directly into the telescope tube, and if firmly screwed home will be both centred on the axis of the tube and correctly aligned to produce a primary image that is also on the axis.

In the case of larger instruments, the O.G. cell is often adjustable. This may be achieved by providing for the cell to be bolted on to a flange on the main tube, usually with three fixing bolts, three lockable screws also being provided to bear against the flange and so permit the cell to be tilted slightly with respect to the flange until the instrument is collimated. An alternative method is a double cell, the outer component of which screws into or is bolted on to the telescope tube, the inner component (bearing the objective) being bolted into it, again with lockable adjusting screws to determine its exact alignment.

Centering the O.G. on the axis is not usually a problem in

refractors of sizes usually used by amateurs; in larger instruments adjusting screws are provided to bear on the rim and so enable the cell to be centred. In all refractors, once the O.G. has been correctly set it should, if left undisturbed, need no further adjustment.

The squaring-on of the O.G. can be carried out in daylight, using a test object such as a brilliantly illuminated ball-bearing; the observer with a little experience, however, will usually prefer to do the job at night, using the more familiar image of a bright star. Once again Polaris will be found to be an ideal object for the purpose.

The test should be conducted on a night when the air is very steady; an assistant will be required, as the adjustment of the O.G. cell cannot be carried out from the eyepiece. The task is made very much simpler if the assistant is also an experienced observer.

The telescope should initially be fitted with a low-power eyepiece, and set so that the star is centred in the field. A higher power eyepiece should then be substituted, one giving a power of × 35 to 45 per inch of O.G. aperture being ideal. The instrument should be focused with great care, to achieve the sharpest possible image of the star. This image should then be examined; if the O.G. is of good quality and is perfectly squared-on it should consist of a very small circular disk, with one or more concentric diffraction rings surrounding it. The innermost diffraction ring is much the brightest, and should be seen in any instrument; the outer rings are much fainter, but one or two of them should be visible with a good O.G.

If the diffraction rings are of uneven intensity, being much more prominent on one side of the image, the O.G. is not perfectly squared on. If they are absent on one side, and diffused into extended 'flare' on the other—thus giving the image an appearance very like that of a comet with a tail—the misalignment is very serious. The direction and amount of tilt needing to be applied to the O.G. cell can be determined by trial and error, one observer watching the image at the eyepiece whilst the other makes slight adjustments to the setting of the O.G. cell. The movement required will be small. For most O.Gs. likely to be encountered by the amateur, and assuming that an inverting, astronomical eyepiece is being used, it will be found necessary to tilt *towards* the eyepiece the side of the objective towards which the flare seems to point. Some observers like to make the final adjustment with the star out of focus, giving a larger image.

The Newtonian reflector. This is the type of telescope favoured by the great majority of amateur observers, largely due to its relative simplicity of construction. A large number of the Newtonian reflectors in use have consequently been made by amateurs, but many of these have been constructed to a high standard and have excellent performance. Unfortunately the excellence of the construction is not, in some cases, matched by the quality of the optical components: no amount of adjustment can make up for inferior optics, however, and so it is assumed here that high-class optical elements have been used.

A Newtonian reflector is properly collimated if the optical axes of the primary mirror and the eyepiece intersect at the centre of the reflecting surface of the inclined flat, and make equal angles with the normal (perpendicular) to that surface. Adjustments are provided in most instruments of this type for movement of the primary mirror and the secondary flat. (It may be mentioned here that a right-angled prism is sometimes used for secondary, instead of an inclined flat; this is somewhat rare, due to the high cost of a suitable prism—perhaps fortunately, for the adjustment of a prism is much more critical, and therefore more difficult, than that of a flat.)

The primary mirror is usually mounted on a three-point support, each supporting stud or screw being adjustable so that the mirror can be tilted. They should also be fitted with locking nuts. In the case of larger instruments screws are sometimes provided to bear on the rim of the mirror cell also, thus giving a means of centering it in the telescope tube.

There are numerous forms of mounting for the secondary flat; there is usually provision for movement of the flat along the axis of the telescope tube, and for rotation of its holder around that axis.

The collimation of a Newtonian reflector is easier with the help of an assistant, and is best carried out in daylight. The telescope is pointed towards the sky so that the mirror is fully illuminated (though not, of course, by direct sunlight!). No eyepiece should be fitted. On looking into the empty draw-tube, the observer will see the flat—possibly not properly positioned with respect to the draw-tube nor 'pointing' along it—and reflected in its surface he should see the bright disk of the primary mirror with a dark spot in it which is the silhouette of the flat-holder. Probably none of these will be concentric, showing that the telescope is not collimated.

The flat-holder should be moved until the centre of the flat is on the axis of the draw-tube; it should then be rotated about the axis of the main telescope until it directs the reflected light from the primary mirror along the draw-tube axis. The edge of the elliptical flat will then appear to be circular and concentric with the edge of the draw-tube.

The next stage of collimation is easier if a high-power eyepiece is fitted into the draw-tube, its lenses having first been carefully removed. (Ensure that you know their position and orientation for subsequent reassembly!) The eye-aperture of the eyepiece will then form a convenient stop, exactly centred on the axis of the draw-tube. If, on looking through this aperture at the flat, the bright reflection of the primary mirror is not concentric with the edge of the flat, the latter is still not precisely adjusted. Further slight rotation of the flat-holder around the telescope axis may be required, or it may need to be tilted so that the angle its surface makes with the axis of the draw-tube more nearly approaches 45°; the adjustment required can easily be discovered by trial and error. Some flat mountings have a screw provided for tilting the flat; in other cases it is necessary to adjust the tilt by altering the fixing to the tube of the 'spider' mounting which carries the flat-holder.

Finally, when the image of the main mirror, the edge of the flat and the draw-tube all appear concentric, the last stage can be tackled. This consists of tilting the main mirror by means of its adjustable three-point suspension, until the dark image of the silhouetted flat appears central in the bright reflection of the main mirror. When this small dark spot is also concentric, the telescope is fully collimated.

The Cassegrain reflector. Neither easy to construct nor cheap, the Cassegrain system is mainly used in the larger observatory instruments; in recent years, however, Cassegrain reflectors have become available in sufficiently small sizes to attract the experienced amateur observer. A combined Newtonian-Cassegrain design is also sometimes encountered.

A Cassegrain reflector is collimated when the axes of its hyperbolic secondary mirror and eyepiece draw-tube are coincident, and the axis of the concave primary mirror passes through the centre of the secondary mirror.

To adjust the Cassegrain reflector it is again helpful to use a high-power eyepiece with the lenses removed. The operation should be carried out in daylight, with the instrument pointed towards the sky to illuminate the main mirror. On looking through the eyepiece aperture the primary mirror should be seen reflected in the secondary, and should be concentric with it. If adjustment is needed, the secondary should be tilted, making use of the adjustment provided in its mounting, until the image of the primary mirror is exactly concentric with the secondary mirror. Attention should then be directed to the small dark silhouette of the secondary mirror, which should be seen exactly at the centre of the image of the primary mirror. If it is not in this position, the primary mirror must be tilted until it is, using the adjustable suspension screws exactly as for a Newtonian reflector. The instrument is then fully collimated.

ADJUSTMENT DURING MAINTENANCE

Even the simplest telescope mounting requires some routine maintenance, and any astronomical telescope will perform more satisfactorily if it is given a periodic check. This can soon be reduced to a regular routine that need not occupy very much time; more often than not little attention will be found necessary, save for the application of a little oil to the moving parts of the mounting. Sometimes the optical components will be seen to require cleaning—this is a task which, if undertaken by the inexperienced, can cause grave and irretrievable damage to valuable components, and should not be attempted without skilled advice and guidance. There are many aspects of maintenance, however, that can be quite safely carried out by the telescope owner himself, provided that he is prepared to be methodical and to exercise scrupulous care.

Most equatorial mountings of any size will not only have axis bearings, etc, that require regular lubrication, but also bearing plates and adjustment bolts which constrain the moving parts to their correct positions and ranges of movement. These must be regularly checked, for should they work loose the moving parts may wander from their ideal positions: this can cause uneven movement or stiffness, excessive and undesirable strain on certain components, and often a loss of accuracy in setting. The telescope owner will soon become familiar with the construction and operation of his particular mounting, and with regular attention of this kind will be able to keep it in good order without difficulty.

Other items which should be regularly checked are the setting circles: the verniers, and possibly the circles themselves, will be provided with a degree of adjustment; the locking screws must be checked regularly, to ensure that they are still tight. Should they be allowed to work loose, it will be necessary to reset the circles as described above.

A major source of periodic trouble is the stiffening of the draw-tubes and rack-focusing mount (if any). Unless these have been deformed by ill-treatment, the trouble will almost certainly be due to the lubricant having become mixed with dirt and then congealed, forming a patchy film of 'goo' which although very thin has a high coefficient of friction. To treat this condition, some fresh paraffin oil or petrol will be required, and some *clean*, fluffless cloth. Fresh lubricant will also be needed: it is most important that this should not be allowed to come into contact with the optical surfaces, and because of its tendency to 'creep', oil is not recommended. For the lubrication of these parts of a telescope the writer prefers to use white petroleum jelly; only a very small amount is required, and a medium-sized jar is sufficient for many treatments.

The draw-tube(s) should be carefully removed from the main tube, and separated; care should be taken to note their order and, especially, which way round they fit. If necessary, tiny marks should be made on them to ensure correct reassembly.

Each portion of tube should then be cleaned, using cloth which has been dipped in paraffin and the excess liquid squeezed out. Only the bearing surfaces should be cleaned; these are normally the outside surfaces of the tubes, but there will also be a collar inside one end of it if the tube in question has another of smaller diameter to be fitted into it. The bearing surface only of this collar should also be cleaned, using a portion of the cloth wrapped round a finger; no liquid must be allowed to touch any other inside surface of the tube. Great care must be exercised to prevent excess liquid being used, and to keep it well away from all optical components. The novice will find paraffin rather easier to handle than petrol.

When the film of 'goo' has been wiped away, the surface should be wiped again, first with a clean cloth and a little paraffin, and then with a clean, dry cloth. This procedure should be repeated on all sections of tube.

Before reassembling the components, the bearing surfaces of each should have a little of the petroleum jelly applied, which should be spread out into a thin film over the entire bearing portion: quite the best implement for this purpose is a *clean* finger!

If the rack-focusing is giving trouble, it may simply require cleaning in the same way; often, however, it will need readjustment of the mounting screws of the plate carrying the pinion shaft, as the tightness of these screws usually governs the freedom of movement of the shaft. Pressure of the plate must be applied evenly on the shaft bearings to ensure smooth movement of the rack and pinion. They will need tightening if the shaft turns too freely, and slight loosening if it is too stiff. It is usually best to dismantle the mechanism in any case, however, to permit thorough cleaning and re-lubrication; the adjustment can then be made during reassembly.

The retaining plate should be removed by undoing the four screws, thus freeing the pinion shaft. Due to the heavy milled knobs at each end the shaft will tend to fall out, and care must be taken to prevent this. The rack strip will now be revealed, screwed to the inner tube. The outer tube may be slotted, enabling the inner tube to be withdrawn with the rack attached; if not, the rack strip must be removed to permit the inner tube to be withdrawn. After cleaning and greasing as described above, the components are reassembled in the reverse order. It is a good idea to apply petroleum jelly generously to the rack strip before replacing the pinion shaft. The four fixing screws at the corners of the retaining plate should be screwed up until they are all 'finger tight'; they should then be further tightened in stages, a quarter-turn *each* at a time. Rotation of the pinion shaft should be checked after each stage, until the right pressure is achieved. This should be sufficient to hold the tube in any position, yet not so tight as to prevent easy movement when desired. This procedure should ensure even pressure, but if the components are old and worn it may be necessary to given an extra quarter-turn to a particular screw in order to achieve even and balanced movement. The final 'tightness' required is best judged when the draw-tube has been replaced in the telescope, with the heaviest eyepiece (and Barlow lens if normally used) in place.

8 : H. E. DALL

Eyepieces for Telescopes and Binoculars

THE DUTCH SPECTACLE-MAKER LIPPERSHEY'S ACCIDENTAL REALISA-tion of the telescope began when he looked through one lens placed near the eye at the image of a distant object formed by another lens of longer focus. This makes clear the reason for calling the smaller lens an eyepiece or ocular. The lens can be either of negative (Galilean), or positive type. The former type gives an erect image because it intercepts the rays from the longer focus lens before they reach their focus or crossing point. The field of view seen is very small because the negative lens cannot con-verge or concentrate the field image into the pupil of the observer's eye. A positive lens eyepiece *can* do this, hence show-ing a much larger field of view.

Thus the function of an eyepiece is to magnify the image formed by the other optical elements of the telescope and to present to the eye a bundle of rays covering a fair viewing angle. It must be coaxial and focused so that the lens of the eye re-focuses the image easily and clearly on the retina.

No optical system is perfect. Imperfections of one kind or another may vitiate the retinal image, but the eyepiece is a good and suitable one if the imperfections are smaller than the imperfections of the human eye itself. If one has to choose an eyepiece for use with an existing or proposed telescope, a compromise is involved in which there are a number of factors:

1. Cost.
2. Correction of optical errors which affect performance, e.g. spherical and chromatic aberration, coma, astigmatism, distortion, ghosts and flare.
3. Efficiency of light transmission.
4. Angle of field of acceptable definition.
5. Eye clearance.
6. Suitability for projection purposes.
7. Hardness and durability of external surfaces and accessibility.

There need be little or no difficulty in the choice of an eyepiece if the telescope is of long focal ratio (focal length/aperture), e.g. f/10 upwards, because for these ratios, the simple 2-element Huyghenian or Ramsden type will perform as well as more complex 3 or 4-element types. Astronomical refractors are usually of long focal ratios, f/15 being common, and they are normally supplied with Huyghenian eyepieces. On the other hand prismatic binoculars are of short focal ratios (f/3 to f/4·5) for which more complex types of eyepieces are essential. The vast majority of reflecting telescopes in use by amateurs have medium focal ratios between f/6 and f/9, giving scope for care in choosing the best eyepiece for the type of observing intended.

The focal ratio referred to is the final focal ratio in the case of compound (eg Cassegrain) telescopes, or those fitted with Barlow or transfer lenses. For example a 6in Newtonian of 42in focus (f/7) fitted with a × 2 Barlow has a final focal ratio of f/14.

Enlarging on the seven factors listed above:

Cost. The fewer the elements in the eyepiece the lower the cost, and if the elements are identical and of cheap glass, so much the better from this one point of view. The most expensive types have five or six elements with several different types of glass of varying degrees of hardness and resistance to weathering. These complex types are usually designed for use on short focal ratio instruments giving wide fields of view and often for use with prisms. If used on f/15 refractors, they give results inferior to good Huyghenian types, except in the size of the field of view.

Correction of optical errors. A lens consisting of a single glass element with spherical surfaces suffers from the two major errors of spherical and chromatic aberration; nevertheless the errors are not severe enough to effect performance in a telescope if the focal ratio is above f/10, implying ray angles not sloping more than about 3° from the axial focal point. The form of the lens and the type of glass affects the severity and it should be understood that a single lens is only suitable for small fields of view.

Eyepieces with 2 elements of similar glass are very commonly used in telescopes and microscopes and have a much larger field of view than the single element type; they are also not generally suited to focal ratios below f/10. To illustrate the effect of these errors and their dependence on focal ratio, an example will be useful.

If a Huyghenian eyepiece is applied to a good f/5 telescope, a halo of aberration will be seen surrounding the image of a bright star or planet. If the aperture of the telescope is then halved by fitting a stop making the focal ratio f/10, the aberration halo will be invisible because it is reduced in diameter by a factor of 8 and by a factor of 64 in area. Thus the visibility of the error can be considered to vary inversely as the fifth power of the f number, an enormously important variation. Replacing the Huyghenian with an orthoscopic eyepiece of good type, the aberration halo at the full aperture f/5 will be no more than that of the Huyghenian at f/10.

Eyepieces of 3 or more elements, made from at least two types of optical glass, are necessary for telescopes of focal ratio f/8 or less if the best performance is required. The only exception to this general rule is when the mirror or object-glass is deliberately or otherwise overcorrected for the major aberrations to match the undercorrection of one particular eyepiece. The two major aberrations, if severe, affect the definition in the centre of the field of view. Other optical errors, e.g. coma, astigmatism, distortion, curvature of field and chromatic differences of magnification, affect only the outer parts of the field. Some attempt is made to reduce these errors in the more complex wide field types of eyepieces, but residual errors are very noticeable outfield, particularly for low focal ratio telescopes and when observing bright stars or planets. The most successful types are those designed and used with particular object glasses and prismatic assemblies. Under these conditions the errors of the various components can be balanced or optimised.

Loss of light in transmission through the eyepiece is inevitably greatest with complex types, but lens coating (blooming) can offset much of this loss, as shown in Fig 28. Every air/glass surface contributes to this loss, and other small losses occur at cemented surfaces and by absorption in the glass. It is not advisable to coat the lens surface nearest to the eye as this usually has to be cleaned frequently and the coated surface is more vulnerable to scratching during cleaning than the uncoated glass.

The reflection loss per air/glass surface is slightly over 4% for crown glass and about 6% for dense flint and dense barium crown. Glasses of the two latter types acquire a natural bloom with age, due to leaching out of the lead and barium in the surface layers in humid atmospheres. This natural state is sometimes beneficial, but eventually the outer surface weathers to a scattering layer of an undesirable kind.

Loss of light by absorption does not usually exceed 1% per element except when poor quality glass is used.

Typical figures for overall loss of light are:

Single-element type   9% uncoated

                                   $6\frac{1}{2}$% coated 1 surface

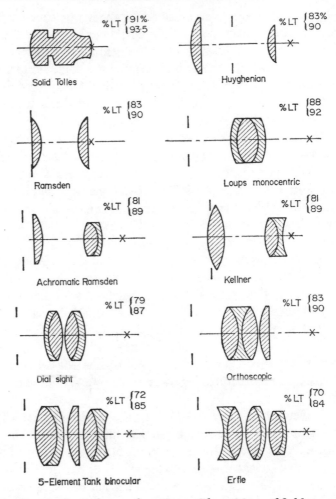

Fig 28   *Ten basic forms of eyepiece with positions of field stop and eyepoint, with percentage light transmission, uncoated and coated (from* Eyepieces *by H. E. Dall)*

2-element type	17% uncoated
	10% coated 3 surfaces
3-element (4-glass/air)	19% uncoated
	11% coated 3 surfaces
4-element (4-glass/air)	21% uncoated
	12% coated 3 surfaces
5-element (6-glass/air)	28% uncoated
	15% coated 5 surfaces
6-element (6-glass/air)	30% uncoated
	16% coated 5 surfaces

The gain from coating in the last item is between a transmission of 70 and 84%. This is a 20% gain in light.

Angle of field. The angle through which the eye moves between an examination of two opposite sides of the field of view is called the 'apparent' angle of view. This must not be confused with the actual field when the eyepiece is used with a particular telescope; for example, if the apparent field is 40° and the magnification given is × 20, the actual field will be about 2°, say four lunar diameters.

Apparent fields less than 35° are considered narrow, 35°–50° are medium, 50°–60° are wide field and 60°–80° are extra wide fields. In no type of eyepiece is definition at the edge of the field as good as in the centre, but the deterioration is more evident in the wider field eyepieces. The edge deterioration is not particularly serious because an object under close scrutiny should be brought near to the centre.

A wide field is useful for searching purposes or for spectacular views of star clusters or wide spreading objects. Vignetting, or partial obstruction by the mounting of the outer ray bundles, is very common in wide and extra wide field eyepieces. This may affect the use in variable star work, but not for general viewing. The majority of telescopes are fitted with medium field eyepieces.

Eye clearance is the distance between the outer lens surface and the exit pupil, i.e. the image of the mirror or object glass formed by the eyepiece. It is here that the various parallel bundles of rays from all points of the field converge and the iris of the eye should be located here if the whole field is to be visible at once. If the clearance is less than 8mm, the iris cannot be placed at the exit pupil and only part of the field of view can be seen without moving the eye about. A clearance of 16mm or more is necessary to see the whole field at once by a spectacle wearer. The clearance is rarely more than 80% of the eyepiece focal length, hence a wide field cannot be seen *in toto* without eye movement with eyepieces less than 20mm focus for spectacle wearers, or less than 10mm focus for others. Thus there is no great value in wide fields for short focus eyepieces; even a 30° field is acceptable for really high powers—especially with a clock driven telescope. For low powers, on the other hand, it is possible (but not common) to have too great an eye clearance. This results in unpleasant shadowing of the field until the eye is withdrawn to the exit pupil. Bearing in mind that the exit pupil is an image of the mirror or object glass formed by the eyepiece, unpleasant shadowing of the field also occurs if the eyepiece has severe spherical aberration for rays in this direction, i.e. with the most nearly parallel rays entering from the side opposite to the eye. The effect can be very disturbing for wide field low power eyepieces of some types, and cannot be eliminated by moving the eye.

Suitability for projection. In general the 3- to 6-element achromatised types are most suitable for photography by projection through the eyepiece. The simple Huyghenian, if correctly made, does not show colour fringes when used visually, because the red and blue virtual images are seen of equal size. When used for projection without filters, colour fringes may be troublesome, but not seriously so for large focal ratios. The Ramsden type also suffers from colour fringes in projection.

Accessibility of the elements for cleaning is quite an important point. In particular the surface next to the eye should not be shrouded with an eyecap which would prevent easy cleaning. The outside field surface is also liable to collect dust, and should be easily accessible for cleaning and yet protected from contact when laid down. The inner surfaces may need attention at, say, yearly intervals and again should be accessible. Permanent burnishing-in of elements preventing dismounting should be condemned. Durability of the glass, particularly of the outer surfaces, is worth consideration. This point is dealt with for specific types in the next section.

SPECIFIC TYPES OF EYEPIECE

Fig 28 illustrates the form and relationship of all the eyepiece types in common use, given in order of increasing complexity. The small cross shows the position of the exit pupil or eyepoint for each type. The position of the field stop is also shown and two figures of light transmission percentage for uncoated and coated lenses is given.

The single element *Tolles* form, sometimes called solid Huyghenian, has a high light transmission and no ghosts, and an overall length about $1\frac{1}{2}$ times the focal length. Spherical aberration is less than one-fifth that of an equal focus Huyghenian, and it gives crisp and colour free images even for f/6 telescopes. The disadvantages are fairly severe in the form of considerable field curvature and no eye clearance and a small field of 28°–30° of which little more than 20° can be seen without eye movement. The focal plane is internal and a groove in this plane forms the field stop. If this is omitted (indeed from any eyepiece) a blurred edge to the field of view results—psychologically less satisfying than a sharp boundary.

Hard crown or barium crown glass is used and the best ratio and separation of the curves vary with the glass type. If the small field is not objectionable, it is an excellent eyepiece for planetary and double star observation.

An achromatic version with a cemented eyecap of dense flint

glass has been used—this puts it into the class of the Loups monocentric with corresponding advantages but higher cost.

The common *Huyghenian* is one of the oldest and cheapest types and well suited to long focal ratios. It is probably more used for refracting telescopes and microscopes than any other kind; both have long focal ratios. Its merits for these duties are considerable. Fields of 40°–50° are usual, flat and colour free, and the lenses are of inexpensive crown glass, durable and not easily scratched. Poor design and manufacture, such as the use of poor non-optical glass, bad centering and wrong separation has often vitiated the performance, but this is not a defect of the type. A real disadvantage is the large amount of spherical aberration, amounting to 3–4% of the focal length for f/10 telescopes and four times this for f/5. The effect of this is described in the previous section on optical errors.

If a telescope is found to be overcorrected (not uncommon in falling temperature gradients), use of this type of eyepiece may well be found to neutralise the error for medium to high powers even for f/6 to f/8 telescopes.

The textbook 3:1 ratio of foci of the elements results in a small eye clearance. A better and much used compromise is a ratio of 2:1 with a separation half the sum of the foci (element foci 0·75 and 1·5 times the eyepiece focus); plano-convex is the standard form of the lenses but variants giving slightly improved results have been used. Another variant with a cemented achromatic eye lens has been used for projection purposes.

The textbook form of the common *Ramsden* eyepiece, consisting of two identical plano-convex lenses of crown glass separated by a distance equal to their focus, suffers from zero eye clearance and in having the field lens in focus together with any dust on its surface. In addition there is a great deal of outfield false colour. The first two defects are reduced by making the separation about 70% of the focal length, which is fairly common practice. This gives an eye clearance of about 20%, not sufficient to enable the field of 40°–50° to be seen without some eye movement. The spherical aberration of the Ramsden is much less than that of the Huyghenian, and it can be used down to about f/6 in short foci.

The field stop is just outside the lens, and the type has been much used for micrometers and surveying instruments which have cross wires coincident with the primary image. For non-spectacle wearers the close eyepoint is less inconvenient, and the outfield colour can be made use of to counteract atmospheric dispersion to some extent for low-altitude astronomical objects. Observers with short sight (myopia) find they still focus dust on the field lens too easily, and get more satisfaction from the Huyghenian, with which this does not occur. The Ramsden is the cheapest 2-element type to produce, and is used in America to a much greater extent than in England.

A cemented *Triplet*, which has been considerably favoured for its crisp image and economy of light, is the *monocentric* or its equally satisfactory variant, the 'Loups' triplet. In the former all the six curves are struck from the same point, the centre being of near spherical crown glass; the two outer lenses are of relatively soft dense flint. With the concentric design, centering of the components becomes of minor importance, but little if any other advantages accrue, and it is noteworthy that many pre-war Zeiss monocentrics did not follow this design, but were of the 'Loups' type as shown in the figure. With only two air-glass surfaces and a good eye clearance of 80% or more, they make an efficient type of eyepiece very free from errors, and there is less curvature of field than the solid Tolles. However, the field of view is small, only 30°. They can be used down to f/5 or even f/4, but are relatively expensive, although the cost of production should obviously be less than the 4-element orthoscopic.

The *achromatic Ramsden* is a great improvement on the simple 2-element Ramsden. A common design has a plano-convex field lens of crown glass and a cemented doublet eye lens, with a plane or slightly convex surface to the dense flint element facing the eye. However, many variations are made with the plane surfaces replaced by convex surfaces. The crown component of the eye doublet is usually of barium crown. This achromatic Ramsden is probably the most common 3-element type extant, and is fitted to many military instruments and to probably 80% of all prismatic binoculars. It has a field from 40°–50°, and an eye clearance of 30–45%. The performance is good and the field flat, and it is usable even down to f/4. The outfield colour of the 2-element Ramsden is eliminated by the corrected eye lens, but the dense flint eye surface is soft and vulnerable to scratches. Although made in such large numbers in the focal range 0·6–1·1in, they are not invariably well made or well designed, although those by well-known makers can usually be relied on. The proper description of this type is achromatic Ramsden, but it is often referred to as *Kellner* type. The eyepiece properly described as of Kellner type is that shown in Fig 28, which has a double convex field lens with a cemented doublet eye lens having a concave surface facing the eye. The type is not common, but had a vogue in microscopy at one time. It has a very wide and flat field (eg 60°), but such a small eye clearance that much movement is required to see the whole field. A further defect is that dust on the field lens is in focus.

Dial Sight Orthoscopic, or *Plossyl*, was used in large numbers in military instruments, and gives a large eye clearance of 80% combined with a good flat field of 40°. The design is also used for rifle sights where a large eye clearance is a vital requirement. It is also used for erecting purposes in gun sighting and other telescopes. It consists of two fully corrected achromats placed crown sides together and nearly touching. Usually the two cemented achromats are of crossed convex form and are identical, but a number of variations in design exist in which the eye doublet is of shorter focus than the other. By this means fields of up to 50° are achieved. Other variants have plane outer surfaces. They are excellent for astronomical use for foci down to 15mm, and the only disadvantage is the vulnerability of the relatively soft eye lens to scratches. The old Browning achromats can be considered as variants of this form, but the separation of the doublet is much greater than for the dial sight type, and is thus like a fully achromatised Ramsden.

True Orthoscopic, to which the name was originally given, is distinguished by the plano-convex eye lens with the plane side next to the eye, and a triplet field lens. This cemented triplet is overcorrected to balance the errors of the simple eye lens, and consists of a double concave dense flint lens between two hard crowns—symmetrical in some designs, but not in the best ones, in which a field of 50° is achieved; the field is usually 35°–40°. The eye clearance is as large as the dial sight type—80% of the focus. It has all the other merits of the dial sight type and is unique for 4-element eyepieces in having all the exposed surfaces of hard and durable glass. There are numerous variations in the pattern, but maintaining the characteristics of plano-convex eye lens. One variation by Goerz, with a plane outer surface to the triplet and a dense barium crown eye lens, achieves a field of 60° at the expense of soft and relatively unstable glass at both ends. A further variant used in some German military instruments has a cemented quadruplet field lens, and a field of 75°–80° is attained with long eye relief. Large fields have also been obtained by using an aspheric eye lens with a Triplet field lens. Another wide field design has a normal eye lens and one aspheric curve on the triplet.

Fig 28 shows a 5-*element* eyepiece used in tank binoculars and giving an extra wide field of 75°. This is selected as one example of the numerous designs of 5-element eyepieces that have been used in military instruments. Most of these designs have six air-glass surfaces, and all suffer to some extent in giving ghosts, which detract from astronomical use. The eye clearance of most designs is 40–50% and their chief merits from an astronomical point of view arise from their wide fields and their availability at reasonable prices in the form of war surplus lenses.

The '*Erfle*' 6-element eyepieces were made in vast numbers in the last war for military use and are still available. They consist of three separated cemented pairs. The eye lens is under-corrected and the field lens overcorrected. Normally fields of 65°–70° are given, but some rather uncommon ones achieve 80°. The most common of this type was made for predictor sights, and has a focal length of 0·75in. For astronomical use it is advisable to beware of scratched eye lenses (made of soft glass), also of deteriorated cement in the three doublets. They are also prone to giving one or more ghosts of bright objects. If in good order they are excellent for wide-field spectacular views and for their suitability for projection. The eye clearance is about 50%. For planetary observation they are not very suitable.

Zoom eyepieces have become available generally from Japanese sources. They have a larger number of separated elements than common eyepieces, and although coated, there is a strong danger in humid climates of scattering of light developing from condensation and filming between the elements, which are not readily accessible for cleaning. For this reason and for the relatively high losses, it is doubtful whether they are advisable for astronomical use, especially as the power range covered is only 2·5:1 and refocusing is desirable after a power change. The alternative of combining a short-focus achromatic Barlow lens with a normal eyepiece of moderately low power can with equal overall length give a power ratio of about 4:1 and the fewer elements are all accessible for cleaning. An equally rapid power change to the Zoom is given by a turret of three or more separate parfocal eyepieces.

BINOCULAR EYEPIECES FOR ASTRONOMICAL DUTIES

The beam-splitter principle as used with high-power binocular microscopes is quite applicable for astronomy, but some loss of light compared with the monocular view is inevitable. This loss is naturally very important in astronomy, where 10% light loss could in general be restored only by enlarging the area of the object glass or mirror by 10%. Some doubt has been expressed as to whether the distribution of a fixed amount of light of low intensity into two eyes reveals as much as the same total of light concentrated in one eye. This is a physiological question not fully investigated. My experience suggests that the binocular view loss is insignificant. The beam-splitter principle as shown in Fig 29, used with high-power binocular microscopes, is quite applicable for astronomical telescopes, particularly if the focal ratio is not lower than f/10 and wide fields are not required.

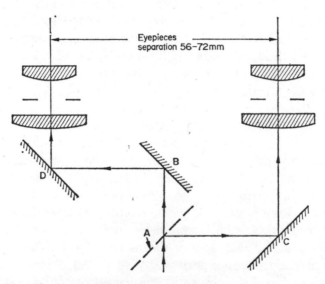

Fig 29 *The beam splitter principle for binocular vision. A is the semi-transparent beam splitter on optical flat or prism. B, C and D are the distributing mirrors or prisms. Unwanted reflections are suppressed by coating*

They are expensive, but can be readily adapted to the telescope. The binocular view of astronomical objects, particular the moon and planets, is enhanced and made comfortable by the use of the two eyes, but it is unlikely to reveal any more details than the monocular view and faint objects will suffer from the inevitable loss of light from the additional optics. However, this loss can be minimised by using dielectric films as beam-splitters (e.g. $\frac{1}{4}$ wave films of titanium dioxide). Precise equality in the beams for each eye is not essential. Even a difference of 2:1 in illumination is barely noticed in practice, and may even be desirable to match eyes of unequal sensitivity. It is far more important that the sum of the two beams is close to 100% of the input, which is the case with dielectric films, but not with metal beam splitters where a figure of 70–75% is more common. This loss is unimportant for microscopy. Multilayer all-dielectric reflecting films (efficiency 98%) can also be used for the other four reflecting surfaces, but they are, and will remain, very expensive until highly automatised methods of producing them are available. Their use for Newtonian flats and Cassegrain secondaries might then be feasible for the amateur.

USING EYEPIECES: High versus Low powers

Beginners in general think in terms of high magnification but experience with high powers soon shows the considerable disadvantages. For most astronomical observing, such as lunar and planetary, experience will show that even when the atmosphere is steady, detail is seen with greater contrast and greater comfort using medium powers. Nevertheless, high powers are desirable for observing double stars or details near the resolving power limit of the telescope, particularly so if the object is bright enough and/or if the observer's eye is itself defective with astigmatism, small opacities or defects of the eye's crystalline lens or cornea. In such cases the exit pupil (perhaps less than 1mm in diameter) is steered automatically into the best area of the eye. By high powers, astronomically speaking, the eyepiece exit pupil (easily calculated by dividing the telescope aperture by the magnification) is less than 1mm diameter; medium powers from 1mm to 3mm and low powers 3mm to 7mm. A first class and experienced eye can see all the detail the telescope is theoretically capable of showing with an exit pupil of 2·5mm, which means a power of ten per inch of aperture or four per centimetre aperture.

The detail is, however, seen with less strain with higher powers, according to the experience of the individual observer.

Fig 30 should make the function of an eyepiece clear. It shows the path followed by two bundles of rays coming from stars on opposite sides of the field of view which after passing through the lens elements converge to the exit pupil at an included angle of 40° where they join the central and other bundles. An eye placed at the exit pupil will see the whole field contained within this angle.

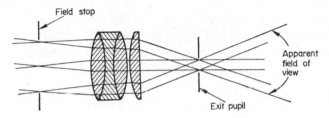

Fig 30   *Paths of three ray bundles from stars at centre and at opposite edges of field of view of eyepiece*

9 : J. C. D. MARSH

Auxiliary Equipment

THE ENJOYMENT OBTAINED FROM OWNING AND USING A TELESCOPE, and the range of observations which can be undertaken, can be extended and enhanced by fitting ancillary instrumentation. The cost need not be excessive, and only a limited amount of mechanical and/or electronic skill is necessary. In this chapter, a number of auxiliary devices are described, any of which can add to the pleasure of the observer and often to the validity and variety of the observations undertaken.

AIDS TO SOLAR OBSERVATION

It cannot be emphasised too strongly that the observer must never look at the Sun directly, either with the naked eye or by means of any optical aid; to do so is to invite permanent damage to the eye. Nevertheless the Sun can be observed simply and easily and with perfect safety. The method is to project the image of the Sun through an eyepiece on to a flat screen (see Fig 31). A choice of eyepiece enables various diameter images to

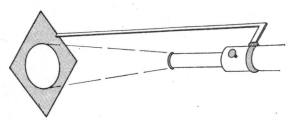

Fig 31   *Projection of the Sun's image through an eyepiece on to a flat screen*

148

be obtained, and focusing is by racking the eyepiece in and out in the usual way. It should be remarked, though, that unless precautions are taken, such as stopping down the aperture of the telescope, the eyepiece might be damaged by excessive heat.

The projection screen can be a simple flat board on to which is pinned a sheet of white card. This method enables various features such as limb darkening, faculæ, granulation and sunspots to be seen and drawn; the rotation of the Sun to be calculated, and, as a long-term observation, the periodicity of the 11·1 year sunspot cycle to be established.

The clearest picture is obtained by fitting a collar around the telescope so as to shade the projection screen from direct sunlight. A slightly more sophisticated apparatus is the sun box shown in Fig 32. The image of the Sun is viewed through a door

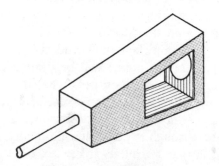

Fig 32   *The 'Sun box'*

in the side of the box, which almost entirely excludes extraneous light. It will be apparent that photographing the solar image with a camera aimed through the door of the sun box is relatively straightforward. For details of exposures etc., see the chapter on astronomical photography.

An alternative method of looking at the Sun is to employ a sun diagonal. This is an unsilvered plane glass wedge which transmits the heat of the sun straight through, but deflects a small proportion of the light image at right angles. Used in conjunction with an eyepiece and a dark suncap good images can be obtained in perfect safety. It should be noted, however, that it is very risky to use a suncap alone with an eyepiece; the heat generated is very likely to cause the suncap to crack, and the intense concentrated light can do irreparable damage to the eye almost instantaneously.

FILTERS

There is an argument that placing a filter in the optical train always degrades the image on the grounds that the light input to the eye is reduced (both by absorption and by restricting the wavelength interval); and since the filter may not be optically flat or truly homogeneous, aberrations will be introduced. While both the above statements are true to some extent they are not the whole story. It is often convenient to use colour filters to emphasise certain markings say, on the planets, or to search for faint colour effects on the Moon. Certainly the overall image brightness will be diminished but—and this is the important point—some parts of the image will be diminished more than the others; that is to say, the contrast (what is sometimes called the signal to noise ratio) will be improved. Thus it will be found that the Red Spot on Jupiter will often appear darker if the planet is viewed through a pale blue filter and lighter if it is viewed through a pale red filter. Likewise, observations of the planets Venus and Mercury by day can often be facilitated by using a pale orange filter which effectively darkens the blue sky and allows the planets to be more clearly seen.

Professional astronomers frequently use filters when making photometric measurements, that is, measurements of the intensity and colours of stars. They use the UBVR system in which the colour U (Ultra-violet) is a band of light centred on 3,500Å; B (Blue) is centred on 4,400Å; V (Visual) is centred on 5,500Å and R (Red) is centred on 7,000Å. An idealised diagram of these responses is shown in Fig 33.

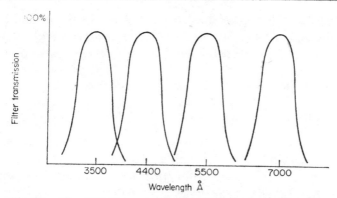

Fig 33 *The UBVR system in the measurement of intensity and colours of stars*

The cost of these filters can be as little or as much as one pleases, depending on the quality and workmanship of the materials used and the accuracy demanded. Useful results can be obtained for a very small sum using one of the Kodak 'Wratten' series, available at most photographers. These filters are squares of celluloid material which may be cut with scissors and mounted easily between plastic or cardboard rings. If they are inserted into the eyepiece mount immediately before the field lens, the degradation of the image is kept to a minimum and the filter can be used with the chosen eyepiece. Fig 34 gives the general idea of the arrangement.

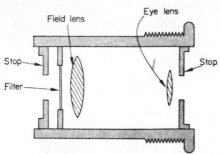

Fig 34 *A filter inserted into the eyepiece mount immediately before the field lens*

A word of warning: if the filter is placed exactly at the focus of the eyepiece, then any dust particles, scratch marks etc, on the surface of the filter will be visible and will noticeably detract from the image.

As well as being of considerable value in visual work, filters may be employed in stellar photography to record images in 'red' and 'blue' light. It can be extremely instructive to photograph the constellation of Orion first with a red filter and then make an identical exposure with a blue filter interposed. Comparison of the black-and-white prints so obtained will show significant differences.

Neutral density filters may also be used if the image, e.g. the Moon under low power, is being observed. These filters have a transmission versus wavelength response which is essentially the same as for the human eye. Consequently the intensity of the image is reduced but the colour balance is unchanged. It is, however, rarely necessary to use these filters, as the majority of astronomical objects are too dim. This kind of filter should never be used alone to reduce the brightness of the solar image, but only as previously mentioned in conjunction with a solar wedge.

MOONBLINK

Much attention has been paid during the past few years to short term events which have been recorded on the moon called Transient Lunar Phenomena, or TLPs. If one of these does occur it may be observed by means of a filter type device known as 'moonblink'. This comprises an instrument which fits, or is held before, the eyepiece and houses two filters, one red and one blue, which are alternately presented in the optical path. Consequently the moon is viewed alternately in red and blue light. If a lunar event in the form of a reddish glow is happening, then this would appear as a dark area in the blue filter and a light area in the red filter. If the switching between the red and blue is carried out in, say half second intervals, the area will blink alternately dark and light and will thus stand out against the normally unvarying lunar background. Various designs of moonblink have been described, the most popular being a rotatable disc containing a pair of red and blue segments (Fig 35). Rotation is by hand. Obviously a more sophisticated version having several filters and being driven by a small motor may be visualised. Another method might be to use an oscillating pendulum driven from a metronome-like device.

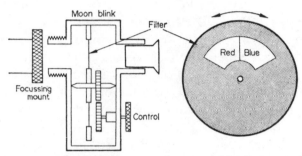

Fig 35 *A moonblink design with a rotatable disc containing a pair of red and blue segments*

POLARIMETRY

A natural step from colour filters is the use of polarising filters. All radiation from celestial objects is electromagnetic in nature, and under certain circumstances the radiation may be polarised, that is, the electric vector of the wave takes a preferred direction. If this is the case then the appearance of the object may be noticeably different if the image is viewed through a polarising filter.

Professional astronomers use a variety of sophisticated techniques which are beyond the scope of this book but the experienced astronomical photographer may make the not too difficult experiment of placing a sheet of polaroid material, obtainable for a very small sum, in front of the camera when photographing a nebula such as the Crab, M1. If an exposure is made with the polarising filter in any arbitrary but known position, and then a second identical exposure made with the polarising filter rotated, say 30° from its original position, the object (if polarised) will produce a modified image. It must be emphasised that this is a delicate observation as in most cases astronomical polarisations are rather small (less than 10%) and instrumental errors could mask true polarisation. However the serious observer might well like to undertake a series of pictures with the polarising filter moved 30° between each exposure.

PHOTOMETRY AND PHOTOMETERS

Photometry is a technique for measuring the intensity of the celestial object either absolutely or relative to another object. The brightness of the object can be measured by using the eye in various visual comparison methods or by using some other light sensitive device such as the photographic plate, photo-transistor or photomultiplier.

Variable star observers compare visually the unknown star with known brightness stars in the same or nearby fields, and this method enables experienced observers to estimate pretty accurately to within 0·1 magnitudes. It is however not difficult to make a photometer which does not rely on comparisons of nearby stars, but on a built-in light-source, the intensity of which is accurately known and a calibration curve drawn.

The simplest of these photometers employs as a light source a bulb supplied from a battery via a variable resistor and a meter. The light from the bulb is diffused by a frosted glass screen and stopped down to a small point of light. This in turn is reflected from a plane silvered mirror into the field of view of the eye-piece where comparison may be made directly with the star under investigation. A schematic diagram of this photometer is given in Fig 36.

For accurate work it is essential that the photometer is calibrated precisely by making measurements of as many known stars as possible. For example, the telescope is pointed to the Pole Star and the intensity of the artificial star adjusted by

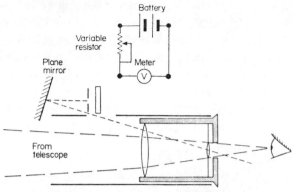

Fig 36 *A simple photometer*

means of the variable resistor until the eye sees real and artificial stars at exactly the same brightness. At this point the reading on the voltmeter is noted, and also the elevation of Polaris. This process is repeated for a number of standard stars covering the apparent magnitude range $0^{m}.0$ to the faintest that can be reliably observed. In this way a number of observations can result:

Star	Magnitude	Elevation	Voltage
Polaris	2·001	40°	3·6

The figure in the magnitude column can be obtained from the list of standard stars in the *Astronomical Ephemeris*. The elevation and voltage figures are those noted at the time of the observation. From this table a calibration curve can be plotted (Fig 37). At first sight this seems an absolutely accurate method, but in practice there are still more factors to be taken into account. The detailed treatment cannot be given here but it will be appreciated that the lower the altitude of the star under observation the greater the light loss due to the earth's atmosphere. An attenuation of about $1^{m}.5$ occurs when the star is less than 15° above the horizon. The colour of the star also affects accuracy as the absorption of the atmosphere is also a function of wavelength, that is, a red star apparently radiating at 6,000Å would appear some 7% brighter than a blue star of the same magnitude radiating at 4,500Å. Useful curves for both extinction and wavelength dependence are given in *The Amateur Astronomer's Handbook*. Finally the measurement depends on the judgement of the observer in determining just when the real star and the artificial star are exactly the same brightness, and so the observation cannot be claimed to be free from personal error.

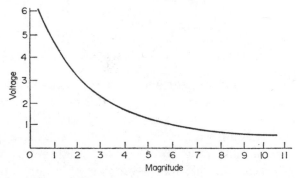

Fig 37   *Calibration curve constructed from elevation and voltage figures*

The rather more ambitious observer wishing to undertake true photo-electric photometry can use one of the less expensive photomultipliers such as the inexpensive 1P21, a DC amplifier to boost the signal and a meter, or ideally a pen recorder to give a tracing of the output signal. Care must be exercised since the photomultiplier requires up to 1,000 volts DC to operate, and well insulated leads and connections are essential for safe operation. It is recommended that such apparatus is only used in an observatory where there will be maximum protection against dampness and the weather generally.

THE POSITION MICROMETER

This is an instrument which is essential for the measurement of small separations between celestial objects such as double stars, the angular diameter of planets, planetary phase etc. The instrument does not measure the absolute position of a single objects, that is, it does not give the celestial co-ordinates, but it does give the position of one object relative to another. Obviously if the absolute position of one of the objects is known, then the co-ordinates of the other object can be found without difficulty.

Many kinds of micrometer have been described and made, and like most ancillary instrumentation, the accuracy of the result depends on the precision of the instrument. Perhaps the simplest kind of micrometer is a grid, capable of rotation, placed in the focal plane of a positive eyepiece. The angular dimensions of the grid must first be found by rotating it until the movement of stars in the field of view in right ascension, that is, with the telescope stationary, is along one line. In Fig 38 this

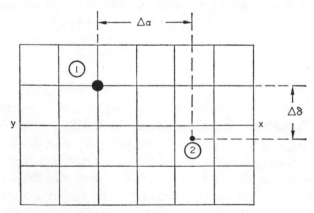

Fig 38   *A grid micrometer*

would appear as a movement from X to Y. Now adjust the declination of the telescope so as to place any star exactly on any of the horizontal grid lines at the extreme right-hand side of the field of view. Allow the star to drift across from X to Y, timing with a stop-watch how long the star takes to go from one square to the next. Knowing that the sky moves at a rate of 15 arcsecs per second of real time, the angular dimensions of the grid can be found. Now, when for example the double stars have been found, align the brighter one (1 in Fig 38) on a horizontal line and allow it to drift along the line until it reaches an intersection with a vertical grid line. Switch on the telescope drive so that the star remains stationary, and measure $\varDelta \alpha$ and $\varDelta \delta$.

A more sophisticated instrument is the filar micrometer, which has a system of cross wires one or more of which is movable; in addition, the whole system is capable of being rotated around the optical axis. Both the axial rotation and the movement of the cross wires is controlled by screws calibrated for a particular eyepiece, and the various displacements can be read off directly from the calibrated scale. For the best results, very fine wires, sometimes spider's webs, are used and in practice it turns out that the wires are extremely difficult to see against a dark field of view. Consequently subdued illumination is essential, usually from a low voltage lamp and a rheostat. The lighting must be indirect and carefully controlled, especially when pairs of faint stars are being measured.

SPECTROSCOPES

Spectroscopy is an extremely specialised branch of astronomy, and it is unlikely that many amateur astronomers can undertake this work. It is however possible to produce spectra with fairly simple equipment, not only of the Sun but of the brighter stars, and to get some idea of the chemical composition of these bodies.

A spectroscope is a device which has the property of splitting up white light into its component colours and displaying these in an array with the well known order red, orange, yellow, green, blue and violet. Also, depending on circumstances, there may be bright or dark lines superimposed on this band of coloured light, or continuous spectrum as it is called.

The essential component of the spectroscope is the dispersing medium, which actually splits the light into its component colours, and which may be either a prism or a grating. Various prisms and arrangements of prisms can be used, but typically the principal section forms an equilateral triangle. If a ray of white light is allowed to fall on one face, the light is refracted through different angles depending on the wavelength, the amount of dispersion depending on the material of the prism. On emerging from the opposite face the light is again refracted and further dispersed. Fig 39 shows the path of the light in such a prism.

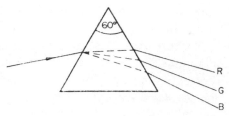

Fig 39 *Path of light through a 60° prism*

A single prism suffers from the defect that the dispersion is not very great, and also that the overall direction of the beam of incident light is changed. This can be overcome by cementing three prisms together, the inner one being made from a different material, thus forming a direct vision spectroscope (Fig 40).

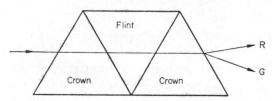

Fig 40 *Three prisms cemented together to form a direct vision spectroscope*

A spectroscope may be screwed directly into the telescope before the eyepiece. Such an instrument has still a rather small dispersion, and although the spectrum of bright stars will be seen, and perhaps a few of the absorption lines, no real measurements can be made.

The simplest instrument which would allow measurements to

be made, and then of course only be photographing the spectrum, is the prism spectroscope, the general arrangement of which is shown in Fig 41. The slit is placed at the focal plane of the telescope which also coincides with the focus of the collimating lens, the purpose of which is to parallel the diverging rays of the incident beam before they enter the prism. The spectrum generated by the prism is photographed by the camera. Since astronomical objects in general are dim even when viewed directly, the spectra produced are even fainter and so relatively long exposures are necessary; several minutes may be considered a minimum. This means of course that an accurate telescope drive is essential, and preferably a guide telescope so that the stellar image is kept accurately aligned on the spectroscope slit.

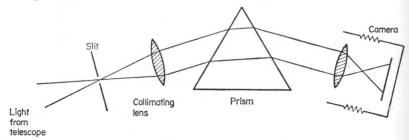

Fig 41   *Idealised prism spectroscope*

Making an efficient spectroscope calls for a degree of mechanical skill. Efficiency depends on the geometrical alignment of the slit jaws parallel to the edges of the prism, the adjustment of the collimating lens to ensure that the beam is perpendicular to the edges of the prism, and the adjustment of the prism for minimum deviation at the required wavelength. Detailed treatment is beyond the scope of this book and the interested reader can find more specialised literature on the subject.

The greatest disadvantage of the prism spectroscope is its inability to transmit light beyond the blue end of the spectrum (limit is about 3,900Å), unless special materials such as quartz are used for the prism. An alternative to the prism instrument is the diffraction grating. This may be considered as a grid of fine wires or lines, accurately parallel, supported and separated by a transparent substrate. Light incident on such a grating is transmitted either straight through or appears as a series of spectra, referred to as different orders. An alternative type of grating is the reflecting grating, which consists of many grooves to the inch of fine lines, ruled in aluminium deposited on a suitable substrate. The incident light is reflected from this grid, and by making the grooves steeper on one side than the other (blazing), concentration of the light into a single order can be achieved. A typical grating can have 15,000 or 30,000 lines to the inch. It is not possible for an amateur to make a diffraction grating, but replica gratings pressed from a master can be obtained quite cheaply and adapted for astronomical use, though with limited success.

The list of instruments discussed in this chapter is by no means exhaustive, and is meant only as a guide to what might be investigated by the serious amateur. Many of the references quoted will lead to even more interesting devices, and contain a wealth of information.

Large Telescopes

IF THE INVENTION OF THE TELESCOPE IS STILL SOMETHING OF A historical problem—for it now seems possible that the Dutch spectacle makers of the seventeenth century, to whom patents were granted, may not have been as early in the field as was once thought—its first serious application to astronomy is in little doubt. In June 1609 Galileo Galilei (1564–1642) heard about an 'eyeglass' that would make distant objects visible, worked out how such a device operated and then, using some spectacle lenses, constructed an instrument for himself.

With his comparatively small instruments, Galileo carried out observations that were epoch-making, showing both the power of the new invention and the inadequacy of previous views of the Universe. But all the same, his instruments had a very restricted field of view and were not of outstanding optical quality. When he observed Saturn he was unable to make out what it was really like, and interpreted the disk and rings as a triple planetary system; yet a couple of years later, when the rings appeared edge on, he saw to his bewilderment that the planet had reverted to a single disk. Not for another forty-three years was Christian Huygens able to solve the mystery, using a large telescope with a focal length of 23ft, and so much freer from optical disadvantages than Galileo's instruments only a sixth as long. Indeed, Huygens' success led to the construction of the first big telescopes with prodigious focal lengths ranging up to over 200ft, and with such an instrument Jean Dominique Cassini

(1625–1712) was able in 1675 to detect a division in the rings that has ever since borne his name.

Long refractors held pride of place for many years, but they were cumbersome. Newton (1642–1727) complained that they were not only difficult to manage, but also shook and sagged so that it was hard to see objects distinctly. As an alternative he believed that lenses should be avoided at least as far as light gathering was concerned, and he decided that the astronomer's salvation lay in using a reflecting telescope. The Scotsman James Gregory (1638–75) had already published proposals for such an instrument, but Newton did not approve of the design and devised his own system in 1668. Meanwhile, in France, a Monsieur Cassegrain (*fl* 1672) designed a reflector of a slightly different kind. Yet opticians had difficulty in constructing good mirrors, which had to be made of metal since there was then no way of coating glass to make it reflect light satisfactorily enough for a telescope, and only in the 1760s were successful reflectors, mainly of the Gregorian type, manufactured by James Short (1710–68). His largest instruments were no more than 18in in length, cost a great deal, and seem to have been used more as objects of pride by the well-to-do than astronomical tools. The first astronomer to build a really big reflecting telescope for purely astronomical use was William Herschel (1738–1822).

A Hanoverian emigrant and one-time professional musician, Herschel obtained royal patronage in 1782, a year after his discovery of the planet Uranus. George III, a fellow Hanoverian, appointed him Royal Astronomer at a salary of £200 which, if not munificent, at least allowed him to follow astronomy as a whole time pursuit. Herschel had long been dissatisfied with commercially available instruments, and as early as 1773 had begun to cast and polish his own mirrors. As his skill improved, so did his ambitions of making a really large instrument, since he was well aware that dimmer stars and greater detail could be observed only with instruments of increasing aperture. In 1781 he tried to cast a mirror of 36in diameter which was to have a focal length of 30ft. The work was carried out in a basement in his house at Bath, but the venture was a disaster, the mould breaking and the molten metal cracking and splintering the stone floor, so that jagged pieces flew in all directions. For a time Herschel had to abandon the idea of so large an instrument, but it was never far from his heart. Later he and his sister Caroline moved to Datchet, near Windsor, and here in 1784 he constructed a Newtonian reflector of 20ft focal length—the dimension by which telescopes were still known at this time. It had a mirror of 18·8in diameter, and performed very well. Indeed Herschel used it for much of the outstanding work he did in what he called 'gauging' the sky—counting and cataloguing stars and everything else he saw.

Amongst the objects he observed were many nebulæ, and these presented a special problem. Some persistently appeared cloudy, but others he could clearly resolve into collections of separate stars, and the question facing him was whether the cloudy ones could also be resolved into stars if he had superior equipment. He had already found that the larger and better the telescope, the more objects that had previously seemed hazy appeared as conglomerations of separate stars. Was it then possible that with a really large instrument, all could be resolved?

When, in 1786, he moved to better conditions at Slough, Herschel determined to try to build a really large instrument. It was to have the unprecedented aperture of 48in and a focal length of 40ft: theoretically it should gather sixty-five times more light than the 20ft reflector or, in more astronomical terms, whereas he could discern stars as dim as magnitude 15 with the 20ft, the 40ft should extend his range to magnitude 17. Moreover its powers to resolve detail should be increased more than two-and-a-half times. But such an instrument would be expensive—there was no question of the mirror being cast at home, and the mounting alone was going to be a huge task. George III was prevailed upon to contribute £4,000, an immense sum in the late eighteenth century, and also provide £200 a year for its running expenses and a further £50 for Caroline Herschel to act as assistant.

Herschel arranged that the metal mirror be cast in London, but the first attempt was not entirely successful, for it was too thin in the centre. A second cracked on cooling, but the third, cast in February 1787, was correct. Meanwhile Herschel had made detailed drawings for the entire instrument and engaged 'none but common workmen' to work under his own supervision. When the first mirror was ground and polished, it took twenty-four workmen at the hands of the grinding and polishing tools to complete it, but it was found that so many men simply could not work together to provide a really smooth motion and Herschel devised and built a polishing and grinding machine which he used on the third mirror. At last, in August 1789, the telescope was ready for test, and came into regular operation a few months later.

The mounting of the 40ft telescope was of wood, the tube being made of iron. The whole structure, built in the form of two triangular wooden pairs of 'ladders', 50ft high, with the tube suspended in between, and a huge wooden base on small wheels that ran on two circular brick walls sunk into the ground, so that the whole instrument could be turned in azimuth. Optically, Herschel adopted his 'front view' method, first tried with the 20ft telescope. Here the main mirror alone is used, being very slightly tilted so that the observer can peer through an eyepiece at the front of the tube directly down to the mirror. In this way the loss of light at a second reflecting surface was avoided. Herschel could move the telescope a small way in altitude and azimuth from his position at the front of the tube, and he could carry out small sweeps of an area of the sky before the whole instrument had to be moved. A speaking tube was fitted so that instructions could go from Herschel to an assistant—usually Caroline—sitting in a small hut at the bottom of the tube. Yet in spite of these facilities, the telescope was not easy to use: the mirror readily became dewed or frosted over, it distorted severely when the tube was at low elevations and, above all, it tarnished quickly and had frequently to be repolished. Nevertheless Herschel did use it on nebulæ and found that although he could resolve some of the previously hazy ones, others obstinately refused to be resolved, and he concluded that there were two kinds of object—star clusters and irresolvable nebulæ. Gradually the telescope, then the largest in the world and a source of wonder to foreign notabilities when they visited England, fell into disuse. In 1839, seventeen years after Herschel's death, it was dismantled. But in Ireland a new giant telescope was to take its place.

William Parsons (1800–67), who became the third Earl of Rosse in 1841, decided to build a giant reflector that would put even Herschel's 40ft in the shade. Like Herschel, he had already had success with smaller telescopes, and now he decided to build an instrument with the unprecedented aperture of 72in. On his estate, a foundry was erected and Rosse experimented with various alloys for the main mirror; in the end three furnaces had to be built so that he could melt sufficient metal in crucibles small enough to be readily handled. Even so it was only at the fifth attempt that a successful speculum was cast. Grinding and polishing was carried out on a machine driven by a steam engine, and the secondary mirror—for Rosse adopted the original Newtonian design—was made of silver, which has a far higher reflectivity than speculum metal. This was a completely new departure, but not so the mounting of the telescope which was less convenient than Herschel's. The tube was suspended between two brick walls, for a wooden support was impossible for so large an instrument, and engineering techniques were not equal to the task of providing an accurate mounting in metal, although the gradual development of the steam engine and associated heavy engineering was to change this within the next two decades.

Nevertheless, in spite of the shortcomings of its mounting and the prevalent cloudiness in Ireland, observations of great importance were made, and it was with this reflector that the spiral nature of certain nebulæ was first observed. In the end, Rosse and his colleague Romney Robinson, Director of the Armagh Observatory, concluded that some nebulæ were really conglomerations of stars, and could be resolved with an even larger instrument. We now know that some 'nebulæ' are gaseous, others star-clusters and galaxies.

While Herschel and Rosse were occupied in constructing huge reflectors, the refractor was changing from an unwieldy instrument of extremely long focus to something more manageable and far more efficient optically. This was due to the development of the achromatic object-glass, most notably by Chester Moor Hall (1703–71) and John Dollond (1706–61). It was a possibility that Newton had rejected because he was of the mistaken opinion that every kind of glass dispersed white light into its separate colours equally, but Hall and Dollond achieved their results using object-glasses with two components, one of crown glass and one of flint glass. However flint glass of good optical quality was difficult to manufacture, as the lead it contained made it hard to obtain an even mixture, and only small lenses of not more than 4in diameter could be made. Not until 1805, when Pierre Guinand (1748–1824), a Swiss optician, devised a method of stirring the molten glass, could large flint glasses be constructed; then sizes crept up to 14in, but it was not until the secrets of Guinand's technique came to England in 1848 during a new revolution in France, that really large lenses began to be manufactured. Lenses of up to 25in diameter were made in Birmingham in 1862 and passed to opticians for figuring, but even so it was in the United States that what were to be the world's largest refractors came to be constructed. This was due to the portrait painter Alvan Clark (1804–87) who, in 1860, turned to full-time optical work and, in partnership with his two sons, founded a company that devised techniques for figuring the largest lenses ever put to regular astronomical use.

In 1884 James Lick, a Californian millionaire, decided to

spend $700,000 to endow a telescope, superior and more power-
ful than any telescope yet made, as one of the many memorials
to his wife and himself. A decision was made to construct a 36in
diameter refractor, and the Clarks were commissioned to make
the object-glass. They ordered their glass blanks from a Paris
glassworks which had bought the Guinand company, and soon
a flint and a crown disk were cast; but the crown was cracked
during packing, and it took three and a half years and twenty
attempts before a second satisfactory crown disk was ready. And,
of course, no such objective was of use without a mounting of
the highest quality which, since the tube of the telescope was to
be 57ft long, promised to be no light undertaking. Tenders were
sought, and in the event the most expensive was accepted because
its design was superior to any of the others. The makers were
Worcester Warner (1846–1929) and Ambrose Swasey (1846–
1937), two engineers who had already taken an interest in
astronomical work and had constructed mountings and rotating
domes. The Lick telescope rightly brought them an immense
reputation. The whole instrument was completed in 1888, and
although one of the Clarks said that 'the only decent thing about
the telescope is the object-glass', it all performed very well and
was used for a vast amount of astronomical research, especially
on double stars and on the determination of stellar line-of-sight
motions. It also showed the practicality of the really big refractor, and when a financial boom swept southern California and
a group of rich men decided to endow a large observatory, it
was a refractor they chose, and Alvan Clark and Warner and
Swasey to whom they turned to construct an instrument of
40in aperture. They ordered the glass blanks from France; but
by the time they arrived, a slump had hit California and the
project had to be abandoned.

It was at this juncture that a well-to-do and enthusiastic young
American astronomer appeared on the scene. George Ellery
Hale (1868–1938), born in Chicago, was attracted to astronomy
in his early teens and had his own small observatory where, in

1889, he developed his spectrohelioscope that was to be so useful in solar astrophysics. In 1892, this private observatory at Kenwood, outside Chicago, was incorporated in the University of Chicago and he became its first director and assistant professor of 'astral physics'. The same year he went to Rochester to attend a meeting of the American Association for the Advancement of Science, and it was here that he heard of the two 40in glass disks waiting to be optically worked—and to be paid for. With Alvan Graham Clark (Alvan Clark's younger son) keen to figure the lenses, and the encouragement of the president of his university, Hale managed to persuade the rich businessman Charles Yerkes to donate money to build 'the largest telescope in the world'. With a mounting by Warner and Swasey, the outward appearance was very similar to the 36in instrument at Lick, but the tube was over 62ft long, the supporting column of the telescope 43ft high, while the object-glass in its mount weighed half a ton and the moving parts that the mounting had to support, some twenty tons. The telescope was in operation in May 1897 and showed itself to be a first-class instrument, especially for the measurement of stellar parallaxes and double stars, and for the first accurate work in stellar photometry.

Hale appreciated the need for and use of large apertures, and with the sophisticated engineering techniques available, he hoped for even larger instruments. But it was clear to him that these must be reflectors, for refractors larger than the Yerkes would require thicker lenses if these were not to sag under their own weight, and they would then absorb too much light. Moreover the problem of casting still larger homogeneous glass disks, free from any defects at all, and then having each of them figured on two surfaces, meant great difficulty and inordinate expense. A mirror had to be figured on one surface only, so its specification was not as stringent as that of a lens. Moreover, since 1851 it had been possible to deposit a silver film on glass, and silver-on-glass astronomical mirrors had been made and used since 1856. As glass was lighter than the speculum metal hitherto used, and easier to figure, very big efficient reflectors became a practical possibility. Hale's father gave him a 60in glass blank, and by 1908, with a grant from the Carnegie Institution in Washington, it was figured and mounted at Hale's new observatory on Mount Wilson, California. But even as the new instrument came into use, Hale's thoughts turned to something even larger, and with the help of George Ritchey (1864–1945), who had a flair for optical and mechanical telescope design, plans were drawn up for a reflector of 100in aperture. Hale then obtained money for the mirror from the industrialist John Hooker, and the Carnegie Institution agreed to provide the mounting.

The 100in mirror was cast in France, but was of seemingly poor quality with many bubbles in the glass; still, as all further attempts to obtain a better disk failed, it was figured by Ritchey and two assistants. The preparation was long and tedious, but in 1919 the instrument—by far and away the largest in the world —began regular work. The design was versatile: optically it could be used at a Newtonian focus near the top of the 40ft lattice-work tube, or with additional mirrors as a Coudé to bring the light to a focus at a fixed point, and this meant that large spectrographic equipment could be employed. The mounting, floated on mercury, was particularly rigid and with the telescope a vast amount of unique work was undertaken. The physical characteristics of stars were studied, and the distances, behaviour and movements of galaxies examined in ways never before possible. Indeed, the success spurred Hale on to the design, finance and construction of an even larger instrument— a 200in reflector.

Discussions about finance, about size, and about design took time and ran into difficulties, but by 1928 the project was under way. A site on Palomar Mountain in San Diego county was finally chosen. Experiments were made for a mirror of quartz (which expands hardly at all with increase of temperature) but these failed, and it was decided to cast a mirror of low expansion Pyrex glass at the Corning Glassworks, New York State. The task was a difficult one, for a solid disk would weigh too much, so a thin ribbed disk was adopted, and then the Pyrex melt had to be held within fine temperature limits during casting so that it did not follow its tendency either to solidify or to boil. By developing new techniques, a successful mirror was cast at the second attempt at the beginning of December 1934. Figuring and polishing had to be conducted carefully without haste after the ten month cooling of the mirror. In November 1947, almost thirteen years later, with the hiatus of the Second World War behind them, the Mount Wilson Observatory optical shops had the mirror ready for use.

The mounting of the telescope was a major engineering feat. The framework tube was carried on two parallel supports, as in the 100in, but in that telescope the cross support at the end prevented observations being made close to the north celestial pole. In the 200in this was overcome by using a horseshoe shaped support at one end, although this had to be constructed in a slightly distorted form so that, under stress in various positions of the telescope, the distortions would compensate for the pressures involved. The astounding result is that the whole 500 tons of moving parts, which float on a thin film of oil, can be pushed round by hand, and to move the entire equipment to compensate for the Earth's rotation, no more than a one-twelfth horsepower electric motor is needed. The movements of the 200in telescope are controlled by a mechanical computer that allows for the differing effects of the Earth's atmosphere (astronomical refraction) with the tube at various positions, and although electronic computers are now common in telescope drives, in 1948 when the 200in was commissioned, this was a great innovation.

The 200in telescope has done all that its founder hoped, and more; its contributions to astronomical research, especially into the far depths of the universe, have been immense. Moreover, spare mirrors of 120in and 98in diameter, produced before the 200in disk itself, have given birth to two new telescopes—the

120in reflector at Lick Observatory and the 98in Isaac Newton telescope at the Royal Greenwich Observatory, Herstmonceux. But with the 200in, telescopes for optical work seem to have come close to a practical maximum, for the Earth's atmosphere imposes limitations on the resolving power and the usefulness of a large mirror. However, the Russians have now completed a telescope of 236in aperture which is to be used at an observatory in the Caucasus near Mount Pastukhov, and with it we may be seeing the biggest of all terrestrial optical telescopes, larger apertures, if ever constructed, being confined to the vacuum conditions of the Moon.

The Russian 236in is unusual in having an altazimuth mounting instead of the now usual equatorial. This makes it easier to provide a very firm support, and although the telescope must be given two motions instead of one if it is continually to follow a star, this now presents no difficulty because the computations needed are carried out by an electronic digital computer. And in this, the telescope is similar to the world's largest 'steerable' radio telescopes—the 250ft diameter parabolic dish at Jodrell Bank, and the new 328ft dish in the Eifel mountains near Bonn, both of which have computer controlled altazimuth mountings.

The advent of the specially designed radio telescope in 1937 and its development since the end of the Second World War has opened a new chapter in the history of the big telescope. Radio waves are some 400,000 times longer than light waves, and so radio telescopes must be very much larger than optical instruments to give a usable resolution: on the other hand they do not need to have their reflecting surfaces made to so close a tolerance as their optical equivalents. Indeed, in the 328ft reflector near Bonn, the surface of the parabola is actually allowed to distort as the dish changes elevation by an average of about 1 millimetre to provide a 'best-fit paraboloid' at every position.

The large steerable radio telescope is an immense engineering undertaking, and when the 250ft Jodrell Bank instrument was constructed between 1955 and 1958, the problems entailed in overcoming wind resistance in so big a structure were novel and challenging, and their solution laid the foundations for the construction of other sizeable paraboloids like the new one near Bonn and the 210ft instrument in New South Wales, Australia. But radio astronomy can also make use of interference techniques in a way that would be unprofitable for the optical astronomer, so that the equivalent of vast apertures can be constructed using two or more smaller radio telescopes and mixing the radio waves from each (interference). In such cases the result is a wave pattern that allows the position of a celestial radio source to be determined with an accuracy similar to that which a single parabola would give if it were as large in diameter as the base line between the small radio telescopes. At the Mullard Radio Astronomy Observatory, Cambridge, the 'aperture synthesis' radio telescope has been developed. Here there is one fixed small radio telescope and one or more movable telescopes to give signals from all over the entire space of the baseline, signals that are synthesised and plotted by computer. At present their largest instrument is equivalent to a 1 mile diameter dish, but a 3 mile diameter telescope is under construction. Moreover, there are plans to use radio telescopes internationally on an interferometer basis, so baselines of hundreds or even thousands of miles are available: possibly the day will even come with an interferometer radio telescope that has one dish on the Moon. Yet whatever the future holds, astronomers are even now using equipment that is a far cry in size and power from Galileo's little 'eyeglass' of three and a half centuries ago.

Photography
at the Telescope

FOLLOWING THE OBJECT—THE TELESCOPE'S DRIVE

EVERYONE KNOWS THAT THE EARTH ROTATES ON ITS AXIS. THIS rotation affects the apparent movement of a heavenly body during the time that will be needed to record its image on a photographic emulsion. Let us start with a few facts:

1. Most telescopes are driven at what is called the sidereal rate. If left to themselves they will follow a fixed star, and make one complete rotation about their polar axes in about 23h and 56m of time as measured by our watches.
2. The Sun loses about 4m per day on the stars, and makes one revolution in about 24h as measured by our watches. Now 4m in 24h is equivalent to 1 part in 360.
3. Relative to the stars, the Moon loses about an hour per day, equivalent to 1 part in 24.
4. The planets may either gain or lose on the stars, but the difference will never be appreciable to the amateur photographer.
5. Most modern telescopes are driven from the mains, generally by a synchronous electric motor (a typical electric clock will also have a synchronous motor). Now all synchronous motors lock themselves to the frequency of the mains supply. If the frequency is that for which the motor is designed (50 cycles per second in Great Britain), the motor will run at the correct speed. But if the frequency drops to, say, 48 cycles per second (and it may well

174

do so temporarily if there is a heavy load on the system) then the motor will run slow by 2 parts in 50, or 1 part in 25.

It will be prudent to assume when designing our photographic telescope drive that an average exposure for the Moon or the planets will need to be about 2s. During this 2s of time the Earth will rotate $2 \times 15 = 30$ seconds of arc about its axis relative to a fixed star.

What will be the effect of the errors referred to above during this time, assuming that the average telescope is being driven at the so-called sidereal rate?

1. If the object is a star or a planet, there will be no noticeable effect, *provided* that the mains frequency is correct. But if by chance the mains frequency has dropped to 48 cycles per second, producing an error of 1 part in 25, then during a 2s exposure the drive will lag by $30/25 = 1·2$ seconds of arc, and the photographic image will be blurred to this extent. This sort of error may be tolerable to a beginner, but it will not be tolerable to the amateur who is striving to take the best photographs possible with his equipment.
2. If the object is the Moon, and the mains frequency is correct, the drive will gain $30/24 = 1·25$ seconds of arc during a 2s exposure, and all the images on the photograph will be blurred to this extent. (Note that a lunar crater one mile in diameter subtends about 1 second of arc at the telescope.) Again, this sort of error will not be acceptable. If the mains are running slow, things will get progressively better, and a mains frequency of 48 cycles per second will produce a lunar rate almost exactly.

All these points indicate that the prudent lunar and planetary photographer should design the drive of his telescope so that its rate can be varied. The amount of variation, assuming that the basic rate is sidereal, will need to be about 1 part in 24 in one direction, to fit the Moon, and an additional 1 part in 24 in the other direction, to allow for a mains frequency of 52 cycles per second (electric clocks have to be speeded up to regain the right time, after they have been allowed to run slow). The total variation needed may be assumed to be 1 part in 10, to be on the safe side.

Note that these awkward errors do not affect solar photography, since the exposures involved are so short. Photographs of the Sun may, in fact, be taken with a telescope which is not being driven at all.

It is possible to make or purchase electronic units which will vary the speed of a synchronous motor. It is also possible to vary the speed of the output from a mains driven synchronous motor mechanically. For some years now the author has used such a mechanical device; it has worked well, and he can recommend it to those who have access to a small lathe.

Fig 42 shows the synchronous motor mounted on an arm which is pivoted about the threaded rod. This arm, carrying the motor with it, can be moved in and out along the rod by turning the adjusting knob. A german-silver cone, which has been lightly

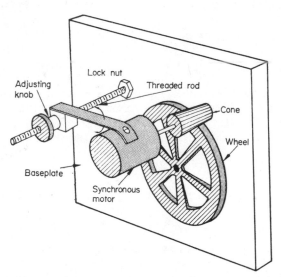

Fig 42 *Synchronous motor mounted on an arm pivoted about threaded rod*

knurled, is mounted on the output shaft of the motor, and rests on the circumference of the wheel, which itself transmits the drive to the telescope in the usual way. The edge of the wheel should be not more than 1/20in thick. The 'apex angle' of the cone is about 3°; this value is not critical. The drive between the cone and the wheel is frictional. In most cases the weight of the motor will be sufficient by itself, but it may be assisted by a light spring if necessary. In this case the fat end of the cone is towards the left of the picture. The speed of the drive may be varied by winding the motor in and out. The author does in fact use two cones, one for 'lunar rate', and one for 'stellar rate', the synchronous motor making one revolution in 8s of time.

OPTICS

Every telescope forms a real image, which can be viewed on a ground glass screen, at its primary focus. The size of this image depends on the focal length of the mirror or objective lens; doubling the focal length doubles the size of the image.

The following formula may be used to equate the size of an image with the focal length of the telescope:

$$\text{Diameter of Primary Image} = F \tan \theta,$$

where F is the focal length of the objective or mirror and θ is the angular diameter of the thing being photographed.

It is not exact,* but it contains no errors that need worry the amateur. It may also be used to determine the equivalent focal length, when the primary image has been magnified. The angular diameters of the Sun, Moon, and all the planets vary from day to day; they may be found in most astronomical handbooks or in any ephemeris. It may not be so easy to find the tangents of very small angles, so they are given overleaf.

---

* The exact formula is:
$$\text{Diameter of Primary Image} = 2F \left(\tan \tfrac{1}{2} \theta \right)$$

NATURAL TANGENTS OF SMALL ANGLES

Angle	Tangent
10″	0·00004848
20″	0.00009696
30″	0·00014544
40″	0·00019393
50″	0·00024241
60″	0·00029089

EXAMPLE

A telescope of 48in focal length is looking at Jupiter, whose diameter is 40 seconds of arc. How big is the primary image?

From the table, tan θ = 0·000194

Therefore diameter of primary image = 48 × 0·000194

$$= 0·0093in$$

The quality of the primary image will depend, among other things, on the type of telescope being used, reflecting telescopes being more satisfactory than refractors. The latter will have to cope with an objective that is not fully achromatic, and the visual and photographic focal planes of the instrument will not be coincident; one may well have to use a filter to overcome these difficulties, and increase the exposures.

The brightness of the image at the primary focus depends on:

1. The diameter of the mirror (i.e. the amount of light collected by it).
2. The focal length of the mirror, and so the size of the image.

The ratio $\dfrac{\text{Focal length of mirror}}{\text{Diameter of mirror}}$ is known as the f number.

At any f number the brightness of the image is the same, regardless of the size of the mirror or its focal length, and therefore the exposures required will be the same. The brightness of an image varies inversely as the square of the f number; thus an image at f/16 will be four times less bright than one at f/8, and will require four times the exposure. Note that it will only be twice as large.

A primary image may be magnified on to a photographic plate or film, very much in the same way as a slide is projected on to a screen. If this is done, all the relationships given above will still hold. The term 'equivalent focal length' replaces the 'primary' focal length of the telescope; the equivalent focal length is equal to the primary focal length multiplied by the magnification. Thus if an f/8 telescope has a focal length of 4ft, and the primary image is magnified 4 times, the new equivalent focal length will be 16ft, the new f number will be f/32, and the magnified image will need 16 times the exposure of the primary image.

Although it is usually desirable to magnify the primary image of the Moon or a planet before photographing it, this process cannot be carried too far without exposure times becoming intolerably long. The f number is always the controlling factor. For the Moon from f/30 to f/60 is reasonable. Jupiter and Saturn will also stand about f/60. Venus will still produce a reasonably bright image at f/80. For the Sun, of course, the limit is governed only by the quality of the magnifying lenses; there is so much light available that it is usually an embarrassment.

PROJECTING THE PRIMARY IMAGE

The primary image of the Moon in a telescope of 4ft focal length will be just under half an inch in diameter. The primary image of Jupiter will be less than one hundredth of an inch in diameter. In most cases, therefore, it will be desirable to magnify this image before imposing it on the photographic emulsion. The amateur may carry out this magnification in one of three ways.

1. **Image Projection using an eyepiece and the camera lens.** In Fig 43 the primary image lies inside the draw-tube. The eyepiece has been adjusted carefully to give a good visual image, in the ordinary way. The rays of light emerging from the eyepiece are therefore parallel. To produce a visual image, they pass through the lens of a human eye, focused at infinity, and on to a human

retina. To produce a photographic image the human eye is replaced by an ordinary camera, with its lens set at infinity, and its stop wide open. The camera should be held square on to the eyepiece, and as close as possible to it, without actually touching.

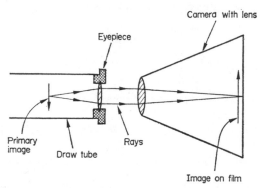

Fig 43 *Image projection using an eyepiece and camera lens*

The size of the projected image in the camera is given by the following formula:

Size of Projected Image

$$= \frac{\text{Size of Primary Image} \times \text{Focal Length of Camera}}{\text{Focal Length of Eyepiece}}$$

Thus a 1in focal length eyepiece working into a 2in focal length camera will produce an image on the film twice as large as the primary image. This method is not recommended for the serious amateur photographer, since it uses more lenses than are necessary.

2. **Image Projection using an eyepiece only.** In Fig 44 the eyepiece, which is a little outside its normal visual focus position, acts as a simple enlarging lens, and projects the image as shown. In the process it will enlarge it in the ratio v/u, and increase the f number and the equivalent focal length of the instrument in the same proportion. Note that no camera, as such, is necessary. All that is needed is something to hold the film or photographic

plate in the correct position relative to the eyepiece, and to screen it from unwanted light. While a single lens reflex camera body without its lens is very suitable, equally good results can be obtained using a home-made box.

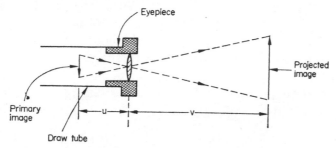

Fig 44 *Image projection using an eyepiece only*

Suitable values of u and v, to obtain any desired magnification m from a lens of focal length f, may be obtained from one or more of the following formulae:

(i) $m = v/u$
(ii) $1/u + 1/v = 1/f$
(iii) $u = f(1 + 1/m)$
(iv) $v = f(1 + m)$

Suppose for instance, that we wish to obtain 4 magnifications using a 1 in eyepiece.
Then from (iii), $u = 1(1 + \frac{1}{4}) = 1\frac{1}{4}$ in,
and from (iv), $v = 1(1 + 4) = 5$ in.
The quality of the projected image using this system can be excellent, provided a good eyepiece is used. Orthoscopics and other multiple lens eyepieces are better than Ramsdens for this purpose; the Huygenian is not likely to give very good results. A good lens from an enlarger should be quite suitable. If there is any choice, the focal length of the eyepiece should be between half an inch and one inch; if in doubt make it a little longer rather than a little shorter. The author has had good results using a $\frac{5}{8}$ in eyepiece to enlarge the primary image ten times (i.e.

to convert f/8 into f/80). The beginner is advised not to attempt more than four or five magnifications to start with.

3. **Image Projection using a Barlow lens.** A good Barlow lens will produce an excellent enlarged image. It should be achromatic, and will then consist of a 2-element 'plano-concave lens', with a negative focal length. It is placed inside the normal focus of the instrument, with its planeside nearer the mirror. It intercepts the converging cone of light from the mirror, and brings it to a focus beyond the primary focal point, as illustrated in Fig 45. The mirror end of the draw-tube is a very convenient point for it.

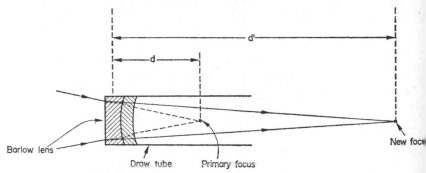

Fig 45   *Image projection using Barlow lens*

If x is the height of the cone whose apex is the primary focus, and y the height of an imaginary cone whose apex is the new focus, then the amplification factor of the Barlow lens is y/x. It is usually referred to as M.

If F = Focal length of main mirror
      F' = Effective focal length of mirror and Barlow combined
      F_b = Focal length of Barlow lens
and d = Distance of Barlow inside the primary focus,

then $M = \dfrac{F'}{F} = \dfrac{F_b}{F_b - d}$

M is equal to one if the Barlow is placed at the primary focus, and infinity if it is placed its own focal length inside the primary focus.

The distance d' from the Barlow itself to the new focus, is given by the expression

$$d' = F_b(M - 1)$$

Suppose that a 4in focal length Barlow lens is placed 3in inside the primary focus of a 6in f/8 instrument.

then $M = \dfrac{F_b}{F_{b-d}} = \dfrac{4}{4 - 3} = 4$

and $d^1 = F_b(M - 1) = 4(4 - 1) = 12\text{in}$

The effective focal length of the instrument will be increased to 16ft, and the f number to f/32.

The secondary mirror of a Cassegrain reflector acts in exactly the same way as a Barlow lens, only by reflection rather than by refraction. Whereas even the best Barlow lens cannot be entirely achromatic, the Cassegrain secondary is. Owners of Cassegrain instruments may therefore count themselves lucky in being the possessors of very high quality astronomical cameras.

A warning. It is not safe to assume that the focal length as described by their makers to eyepieces are correct. For instance an eyepiece said to have a focal length of 1in may well have a focal length anywhere between 0·8 and 1·2in. For this reason the actual magnification obtained when a telescope camera has been designed and made should always be checked by observation. The procedure is as follows:

(i) Set up the equipment and take a picture of a reasonably large and easy object. The Moon and Jupiter are very suitable. Note the time.

(ii) Develop the negative, place it in an enlarger set to a known magnification (say × 10) and measure the diameter of the image. Work out the diameter of the image on the negative.

Then,

Equivalent focal length of telescope

$$= \frac{\text{Diameter of image on negative}}{\tan \theta}$$

Where θ = the angular diameter of the object being photographed, and

$$\text{Magnification} = \frac{\text{Equivalent focal length}}{\text{Primary focal length}}$$

Knowing the magnification, the f number of the camera system and the focal length of the eyepiece can be obtained from the formulæ set out in the preceeding sections.

PHOTOGRAPHY OF THE MOON AND PLANETS.
TAKING THE PICTURE

In addition to the telescope and a good quality eyepiece or Barlow lens, three things are needed before a photograph of the Moon or a planet can be taken:

(i) A plate or film, coated with light sensitive emulsion.
(ii) Something (not necessarily a commercial camera) to hold the film in the correct position.
(iii) A shutter which will give exposures ranging from between about 1/30 and about 4s.

Emulsions, plates and film. Emulsions, as well as films and plates, come in various types and sizes. They all consist of minute grains of a silver salt embedded in gelatine, and mounted on glass or film. They may be likened to the materials to be found on the sea shore. The grains in a fast emulsion (say above 150ASA) are gravel; they respond to faint light, but lack resolving power, and may well become visible if the negative image is enlarged more than about 4 times. A medium speed emulsion (50 to 150ASA) may be likened to coarse sand, and a slow emulsion (20ASA) to fine sand. The slow one needs a lot of light to activate it properly, but it stands enlargement well, and its resolving power is comparatively very high.

Each observer must find his own favourite emulsion by experiment. There is little to choose between the various makes on the market. The author has used Kodak and Ilford 35mm film in the 120ASA speed range for the Moon and planets at f numbers varying from f/60 to f/80, and similar film in the 50ASA speed range at f numbers around f/40; the f numbers are chosen to make the exposures the same for the two types of film. So far as the Moon is concerned, he is still constantly returning to his first love—an Ilford orthochromatic plate rated at 10ASA and used at f/30 to obtain a picture of over half the Moon on one negative.

The beginner should start by taking portraits of the Moon, and will be unwise to graduate to the planets until he is satisfied that he is the master of his photographic emulsion and its processing. If an orthochromatic emulsion is used to start with, it may be developed in a dim red light; it is a great asset to be able to see just what is going on during development, and to be able to adjust the procedure accordingly. The user of a panchromatic emulsion (which is prudently developed in total darkness) will not know when the process is complete and it is too late to do anything about it, whether the lack of any image at all is due to underexposure or to some other cause.

The telescope 'camera'. The body of a single lens reflex camera is very convenient indeed for photographing the Moon and the planets, using eyepiece projection or a Barlow lens. Not only does it provide something to hold the film in the correct position, but also a shutter and a very convenient means of making certain that the image on the film is in the right place and is in focus. However, the author cannot stress too much that you do not need an expensive camera body to take good astronomical photos. Fig 46 shows the arrangement which the author has found very convenient for use with his 12in Newtonian reflector, in which a stout beam replaces the more usual tube. The camera body may be moved in or out along the aluminium supporting tube, to vary the magnification. Note the small holes drilled in

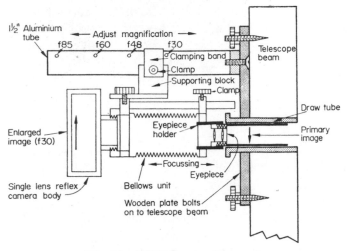

Fig 46 *The telescope 'camera'*

the tube, to indicate the settings for various convenient f numbers. These f numbers increase in the ratio of 1·414 (√2); at f/60 the exposure will need to be twice that at f/48, which will itself need to be twice that at f/30. The f/85 setting is reserved normally for Venus. Once the camera body has been set to the correct position, the primary image is brought to a focus in the viewer by winding the bellows unit, which carries the eyepiece holder and the eyepiece, in or out. This unit is easy to use, and has given very good results. However it is expensive to buy and the camera body and bellows are not easy to make.

Fig 47 shows a camera box which may be made at home for a few shillings; the author has used one for many years and it gives excellent results. The box, which may be made of wood or metal, bolts on to the telescope beam or tube. The eyepiece is fitted to the draw tube in the usual way. A door in one side of the box allows a hand to be inserted for focussing. The author straps a quarter-plate holder to the back of the box; there is no reason why an old camera back should not be used instead. Arrangements must be made to enable the observer to look right through the back of the plate holder, or camera body, so that he may focus the image, and the hole needed must be light tight as and when required. The author draws both slides of his plate-holder, and then examines the image on a special focusing plate through a cheap × 10 pocket magnifier, which is held on an arm so that it focuses precisely in the correct plane. The focusing plate is an ordinary plate which has been fixed in hypo; if a small square of the emulsion is scraped away, the image will be viewed through clear glass, and the 'cliff edges' of the emulsion round the clear window will be very prominent, and will help to bring the image exactly into the plane of the sensitive surface of the plate. Similar arrangements may be made if an old camera body is used with film.

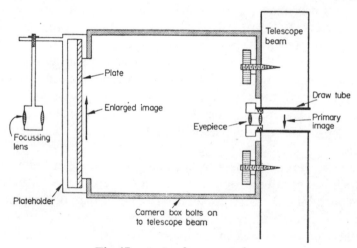

Fig 47 *A simple camera box*

Whereas the owner of a single lens reflex camera will be able to watch the object he is photographing until the instant of exposure, the astronomer using a camera box will have to work blind once he has inserted his film or plate into the holder, since the shutter will have to be shut, and there will be no 'through view'. For this reason it is essential to have a well adjusted finder fitted to the telescope. Nothing elaborate is needed (the author's finder has a 1½in diameter 10in focal length objective), but it should be well mounted, and have a good graticule. Time spent in adjusting this finder will never be wasted.

Shutters. The owner of a single lens reflex camera body will also own a first-class shutter. He will need to take care that the slam of the mirror as the shutter operates does not shake his camera at the vital moment. As is the case so often in telescopic practice, sheer mass is an asset; so do not make your camera fittings too light and flimsy. The author has never been troubled with mirror slam while using the equipment shown in Fig 46.

The camera box in Fig 47 has no shutter, so something must be bought or made to use with it. Old camera shutters are very satisfactory, but care must be taken to make sure that they do not mask some of the light rays coming to a focus from the main mirror. If there is any doubt as to the size of shutter required, point the telescope at the full Moon, and trace round the edges of the 'image' formed on a sheet of paper placed where the shutter will eventually be. Old hats make excellent shutters, but most observers will need three hands to use them successfully.

Fig 48 shows a rotary shutter that can be made at home, and which has been used by the author for several years. Although designed for use with a tubeless telescope, it may be adapted to work with almost any instrument. It may be placed at any convenient position between the main mirror and the photographic emulsion. Although it is not completely light-tight, it is perfectly satisfactory if used after dark. The shutter consists of two light cardboard or metal leaves of the shape shown (part of one of the leaves is shaded, the whole of the other may be seen in the figure); they are mounted concentrically on the spindle, and may be locked in any position relative to one another, so as to vary the size of the gap. Here the gap is 90°, giving a 'half value' exposure; the greatest gap is 180°, and the least gap nothing at

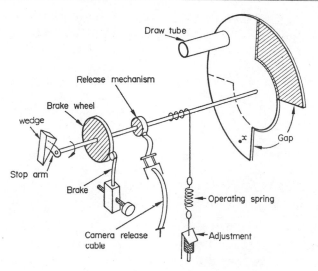

Fig 48   *A simple to make rotary shutter*

all. If this shutter is adjusted to make one half rotation in, say, 2s, the exposure may be varied between 2s and a fraction of a second, by adjusting the gap. To set the shutter the leaves are rotated anti-clockwise, until the point x comes opposite the end of the draw-tube. This tensions the operating spring, and sets the release mechanism. When the camera release cable is operated, the shutter rotates clockwise, and the exposure is made during the time the gap allows light to pass into the draw-tube. The stop arm and wedge prevent the shutter from making much more than one half revolution. Since the shutter starts to rotate before the exposure is made, and terminates the exposure before it stops, there can be no vibration. The brake and brake wheel make the rotation of the shutter more even, and counteract any small out of balance effects which may be present. The brake shoe is mounted on a strip of light spring steel, or similar material.

Exposure times. There is only one infallible rule concerning exposure times in astronomical photography, and this is that they should be as short as possible. It is advisable to proceed as follows:

(i) Think of an exposure value, which may perhaps be based on the points made below.
(ii) Take one picture using this value. Take another using twice the value, and yet another using half, all on the same film or plate.
(iii) Develop the film, and see which exposure is best.
(iv) If none of them produces a good image, decide whether you gave too much or too little exposure, and apply the same process to a second mean value.

If this is done the best exposure will be obtained quite quickly.

Do not forget that exposure times can be affected by development procedures (one can easily give one quarter the exposure recommended by the manufacturers and get a good image, by developing for longer than normal). Above all, do not be afraid to experiment.

What is the correct exposure value? The answer is that it will vary with f number, ASA number (film speed), development procedures, and the weather! However, the following table may be useful. It assumes that the sky is clear, and that the development techniques referred to later are followed. If ordinary commercial development must be used, then the exposure values given here should be quadrupled.

For ASA 50 film and an f number of f/42, use the following exposure times.

Venus	1/60	or 1/30s
Mars	1	or 2s
Jupiter	1	or 2s
Saturn	2	or 3s
Moon: age 4 days or 25 days		3s
8 days or 21 days		2s
12 days or 17 days		1s
15 days (full)		$\frac{1}{2}$s

The Moon values should produce reasonable detail near the terminator; the limb will be overexposed.

Exposures required vary inversely as the ASA number; for example, for a film of 100ASA, the exposures above may be halved. If f/42 needs a one second exposure, then f/30 needs half a second, f/15 one-eighth second, f/60 two seconds, and f/85 four seconds. Remember that the exposure required varies as the square of the f number.

DEVELOPING ASTRONOMICAL PLATES AND FILM

If the amateur astronomical photographer is to make much progress, then he must develop his own film. This is not difficult, and does not require the use of a darkroom if tank development is used.

Processing techniques for astronomical work are basically the same as those for ordinary work; they can be found in any good book on photography.

It is in the development procedures that the astronomer can vary standard techniques to advantage. Once again to experiment is to obey the golden rule—vary the strength of the de-

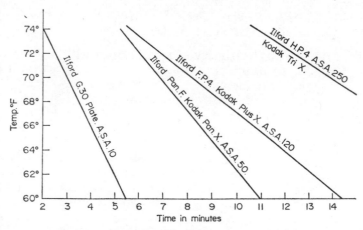

Fig 49 *Development times in Acutol* 1 + 10

veloper, vary the temperature and the time of development, and see whether the results are better or not. Above all, never be satisfied that you have found a perfect technique.

Fig 49 gives the development times found to be reasonable by the author, using Acutol developer, after several years of experiment. They are *not* infallible, and should only be used as a guide. At the present time the author prefers to develop at a temperature of about 72°. This is much higher than the normally recommended temperature of 68°, but it does seem to produce less grain in the finished product.

The methods outlined above 'increase' the speed of a film about four times above the manufacturer's recommended value, for example they make it possible to use Pan F as though its speed were 200ASA instead of 50ASA. There is no noticeable increase in the grain size of the film.

PHOTOGRAPHING THE SUN AND THE STARS

It is not the author's intention to discuss photography of the Sun here, partly because there is not space and partly because specialised telescopes are required to do it properly.

Stellar photography could have a chapter all to itself, but a few guidelines may be given. So when there is no Moon, and the planets are all somewhere else, proceed as follows:

(i) Strap your ordinary camera, loaded with fast colour film, on to the back of your telescope, and point it at the Great Nebula in Orion's sword for about 5m, guiding as carefully as you can.

or (ii) Put your ordinary camera on a stand, and give the same nebula a 3s exposure, using fast colour film.

or (iii) Load your camera with HP4 or Tri-X film, mount it on a stand, and give any group of stars (the Pleiades can be recommended) a 3s exposure. Develop this film in half-strength developer (eg Acutol 1 + 20) for one hour, or even longer if you feel brave.

Plate 1 German mounting, skeleton tube, with counter weights, bearing Henry Brinton's 12in reflector

Plate 2 **(above)** The 82in McDonald reflector shows g design in an English mounti Both piers are massive, while polar axis is sturdy and tapere from the attachment point of the declination axis, which i itself stubby and rigid. The counterweight is positioned of the way, and its weight be near one of the p.a. supports lessening the strain; **(left)** the classic example of the Yoke the 100in Hooker reflector a Wilson. Visibility of the Pol and sub-polar area was sacri for the great stability offeree this design, for what was the the largest telescope in the world (Photo from the Hale Observatories)

Plate 3 **(top)** Terence Moseley's 14.6in Newtonian on a Fork mount, which was relatively easy to build. The large worm-wheel is positioned between the brass polar axis bearings, which are mounted on steel supports bolted to a heavy concrete block. The declination axes are in self-aligning bearings, and carry a slow-motion and setting circle. The tube is completely rotatable

(Bottom) This view of Moseley's mount shows the p.a. rotating in brass bearings, seated in aluminium housings. These are bolted to welded steel supports on a base-plate which is itself bolted to a concrete base block. The bronze wormwheel incorporates a slipping clutch, friction being regulated by the two springs. The large R.A. circle is mounted above the upper bearing, partly hiding it

Plate 4 **(top)** Horseshoe. Currently the ultimate mounting for large reflectors, it was designed by Porter specially for the 200in Reflector. Note the massiveness of the whole mounting. However, although the moving parts of the instrument weigh 600 tons, it is easily driven on its almost frictionless bearings by a 1/12 hp motor (Photo from the Hale Observatories)

(Bottom) Pillar mounting for a 4in refractor, covered by an 'umbrella'. Not ideal, but better than nothing. The whole 'umbrella' simply lifts off

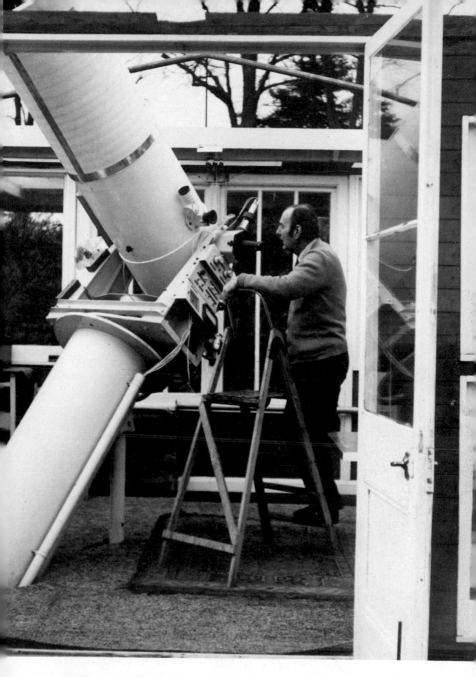

Plate 5 H. K. Robin with his reflector at Tunbridge Wells

Plate 6 Patrick Moore's 5in Cooke refractor at Selsey, on a German mounting

Plate 7 **(top)** Peter Gill's solar projection box, attached to a 3in refractor; **(bottom)** projecting the Sun. A screen has been fitted over the end of a 4in refractor

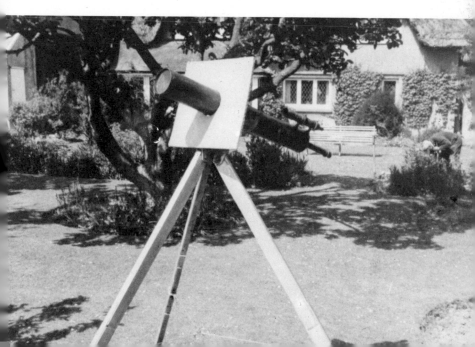

Plate 8 **(top)** Henry Brinton's run-off shed at Selsey, constructed in one piece; **(bottom)** the rails-and-wheels arrangement for Patrick Moore's run-off shed; a two-section construction covering a 12½ in reflector

Plate 9 **(top)** Run-off shed covering Patrick Moore's 12½ in reflector at Selsey; **(bottom)** the run-off shed closed

Plate 10 (top) Patrick
Moore's run-off roof
observatory, covering a
5in refractor. In this
photograph the roof is in
its 'off' position and the
telescope can be seen;
(bottom) chain-and-
handle arrangement for
winding back the roof of
the observatory

Plate 11   (top) J. Hedley Robinson's run-off roof observatory at Teign-mouth. The roof slides back in two sections; **(bottom)** Robin's run-off roof observatory with sliding roof

Plate 12 (top) Dome at Selsey covering Patrick Moore's 8½ in reflector;
(below) rotatable section of this dome, showing the manner in which the
slit opens

Plate 13 **(top)** Overall view of Patrick Moore's dome; **(below)** 'brake' on the rail to stop the top part of the dome revolving in a high wind

Plate 14   (top) Dome at Teignmouth, formerly used by J. Hedley Robinson to cover his 10in reflector. The entire dome rotated—not merely the upper section. The arrangement was not entirely satisfactory, and has now been replaced by the observatory shown below

Plate 15 Commander H. R. Hatfield's dome at Sevenoaks, over his 12in photographic reflector. This is aptly named the 'Beehive' Dome

Plate 16 (top
Hedley Robin-
son's dome wi
shutter open;
(below)
A. C. Curtis's
dome at Win-
chester

12 : L. WILSON
Cinéphotography

INTRODUCTION

THE WORD CINÉPHOTOGRAPHY GENERALLY CALLS TO MIND THE projection, in rapid succession, of a number of photographs to produce the illusion of continuous movement, this process being possible because of the effect called 'persistence of vision', whereby the human eye retains each image for a short but finite time. Of course, in the normal use of cinéphotography, the successive photographs are taken, when the film is made, at the same rate as that at which they are eventually to be projected. In astronomy, however, we can distinguish two main groups of applications of multi-exposure photography which might reasonably be called cinéphotography, but which differ in that in the first group the essential point is that the negatives be taken rapidly and in the second that the final film be projected rapidly.

The first group includes all those investigations where it is necessary to follow some event which occurs rapidly by human standards—examples are given in the first column of the table. When the negatives have been developed and printed we may wish to run through the film at a somewhat reduced speed to follow the intricate changes recorded, but it is more likely that we shall want to study or measure each frame at leisure. It should perhaps be explained that the phrase 'quickly by human standards' does not necessarily refer to events taking very much less than one second, but simply to events that change before the eye can make a quantitative measurement of position or brightness.

Film taken quickly	*Film projected quickly*
Lunar surface profiles	Solar surface phenomena
Details of meteor trains	Rotation of the Sun
Occulation timing	Details of the Jovian atmosphere
Atmospheric turbulence	Phases of the Moon and Venus
	Variable stars
	Motion of comets
	Motion of satellites of Jupiter

The second group of applications really involves the process referred to as 'time-lapse photography'. In many branches of astronomy, events occur so slowly that it is very difficult for the brain to relate developments into any meaningful pattern. As an everyday example, consider that familiar device the traffic light. The sequence of colours shown may take, say, forty seconds: this may not seem long, but if you had never seen a traffic light before you would have to watch one for two or three cycles of colours before you could be sure you had established the pattern involved, thus requiring about two minutes. Yet by taking photographs just sufficiently often to record all combinations of colours which occurred, say one every two seconds, and projecting the film at three or four frames per second, you could establish the pattern in fifteen or twenty seconds. It is this principle that is applied in the examples given in the second column of the table.

TECHNIQUES

In considering the techniques of astronomical cinéphotography, we must again look at the two groups of applications mentioned above. Let us take the second group first. Here we are not concerned with making exposures quickly: we shall rarely need to make more than one exposure in five minutes, so there will be plenty of time to make long exposures if not much light is available. Any type of telescope-camera combination that might normally be used for astronomical photography will do, subject to the usual conditions that exposures of more than about one or two seconds will probably require that the telescope be driven in Right Ascension, and that small images may require the use of a Barlow lens to enlarge them to reasonable proportions. If the image is enlarged in this way the exposure required will also increase, and a happy mean must be found. No more need be said here about the photographic aspects of taking the original negatives: the reader is referred to standard texts on solar, planetary and stellar photography.

However, it must be remembered that these negatives are to be printed and assembled to form a ciné film, and if the final film is to be a faithful record of the order and relative durations of the events recorded, the negatives must be made at reasonably regular intervals. It would clearly be ideal if they could be made on standard 8, 16 or 35mm film suitable for the projector to be used; otherwise some means must be found of printing from the negatives on to standard film. This will certainly involve making a special light-proof carrier for the film, and may also require the use of an enlarger if the original negatives are very large or very small.

Turning now to the first group of applications, we shall find that the remarks made above generally still apply, but in addition we shall have to arrange that the film be advanced automatically between exposures: for the exposures will probably be made at intervals of between ten seconds and one-tenth of a second, and there will be no time to change plates or wind film on by hand. Automatic ciné cameras that can make exposures at this sort of rate are commercially available, but the cost of such equipment will preclude its purchase by most amateurs. Yet the complexity of such a device is not so great as to put its construction beyond the capabilities of the amateur astronomer who has already built, say, an equatorial reflector with motor drive. The basic features are a light-proof box to contain the film, which is held on spools before and after it passes through a frame clamped in the focal plane of the system; a polished, spring loaded pressure plate to hold the film against the frame so that the image is in focus on it; a motor driven sprocket wheel to advance the film, and a suitable shutter. An examination of the film advance mechanism in the body of an ordinary 35mm camera will illustrate most of these features; if the film is to be advanced quickly some slack should be allowed on both sides of the drive to avoid snapping the film. For most purposes a shutter can be made by cutting a slot in an opaque disk which is rotated in the optical path in front of the film.

Clearly there is great scope for ingenuity here and the kind of system devised will depend to a great extent on the actual application. As a good example of what can be done, see reference (3). A major requirement is likely to be the need to know the exact time of one of the exposures in a series, and some suggestions as to how this may be done in particular cases are given later.

APPLICATIONS

The particular applications mentioned in the table will now be dealt with in turn. Those marked with an asterisk are likely to contribute something new to our knowledge of astronomy, whereas those not so marked are for interest only.

*Lunar surface profiles. As the sun sets over a high lunar peak, the tip of the shadow of the peak travels across the landscape still illuminated. By measuring the position of the tip of the shadow as a function of time it is possible to calculate the vertical height of places the shadow crosses below the top of the mountain, and so build up a profile of the terrain. A complete summary of the important points of the method will be found in references (3), (4) and (5) and they will not be repeated here. It must be noted that a substantial telescope—of probably at least 10in aperture—would be needed for this work. Exposures would have to be taken at about half minute intervals and the time of each exposure must be known. Provided the film advance and

exposure tripping mechanism operate smoothly, it will be sufficiently accurate to start a stop-watch at the instant the shutter
is heard to operate and stop the watch at a noted time given by
the telephone speaking clock. Subtraction of the time shown on
the watch from the time given by the clock will yield the instant
of exposure. This can be repeated at intervals during the running
of the camera and the times of intermediate exposures found by
simple interpolation.

Although this work is still of value at the time of writing, it is
to be expected that relative height determinations of much
greater accuracy will soon be made from the stereoscopic pairs
of pictures taken by the American probes in low orbit around
the Moon.

*Details of meteor trains. The problem of photographing meteor
trails using long time exposures is well known: some hours of
exposures are often needed to give a reasonable chance of
securing a good image, though the time may be reduced by using
a wide-angle camera and suitable film. In addition, the number
of meteors available is often three or four times its usual value
during the maximum of an important meteor shower. Even so,
much film usually has to be sacrificed to secure useful results:
cinéphotography is clearly of no use in this context.

However, of even more interest is the possibility of recording
details and changes in the glowing trains of ionised air that many
bright meteors leave behind. This project is probably the most
difficult mentioned in this chapter, for it will be necessary to
construct or obtain not only a very rigid ciné camera capable of
taking at least two or three frames per second, but also to construct a light-sensitive switching device to start the camera when
a meteor appears.

It is this latter device which will be most difficult to construct
unless the builder has some knowledge of electronics. A suitable
circuit is given below in Fig 50: the relay could also be used to
operate the shutter of an objective prism spectroscope for ob-

taining meteor spectra provided the relay operates quickly enough, or to sound a buzzer to inform the operator that the camera has been activated. It should not be overlooked that the circuit shown only switches the camera on: it must be stopped, and the circuit reset, by closing switch 'B' for a moment.

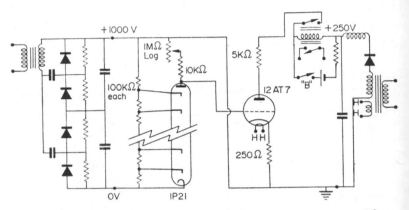

Fig 50   *A light-sensitive switching device to start a camera. The 250 and 1,000V power supplies are only one possibility. Any simple PM tube like the 1P21 will do; only one half of the 12AT7 double triode is used. Use a 2-pole changeover 10mA relay, connecting the 250V HT through the Normally Open contacts on one pole as shown, and the load (motor, etc) through the Normally Closed contacts on the other pole*

An image of that part of the sky that the camera covers must be projected on to the photocathode of the photomultiplier with a suitable lens or mirror. This need not be of high quality, but the greater its diameter the better as the circuit then becomes more sensitive. Reference (7) may be consulted for comments on constructing the circuitry, and great care must be taken to avoid all possibility of shocks from the photomultiplier high tension.

*Occultation timing. It is still a matter of considerable importance in astronomy to know with the maximum accuracy the instant when a star is occulted by the Moon. The equipment described in this section could be used to determine the time when a bright star disappeared behind, or reappeared from behind, the dark limb of the Moon: for reasons that will become clear it could not be used for disappearances or reappearances at the bright limb.

The essential features are a ciné camera in which the film advances steadily at a rate of about 1 centimetre per second attached to a telescope of at least 4in aperture, a plate just in front of the focal plane which is shaped to obscure the image of the illuminated part of the Moon, a rigid equatorial mounting for the telescope with a good drive and a simple circuit activated by a microphone to light a lamp quickly when it 'hears' the last of the three 'pips' of a time signal from the telephone speaking clock. The film in the camera is exposed continuously—there are no shutters and the film advance never pauses. The primary focus image produced by the telescope should normally be used. Light from the lamp which provides the time indication must be introduced into the optical train in the form of a parallel and slightly off axis beam from a slit immediately in front of the lamp. The telescope will then produce an image of the slit near to, but not overlapping, that of the star. Fig 51 shows how this can be done; note that the image of the slit on the film should

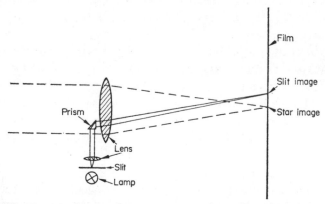

Fig 51 *Size and position of the prism must be such as to obscure as small an area as possible of the main lens*

have its largest extent at right angles to the direction of film advance. A circuit which can be used to light the lamp is shown in Fig 52.

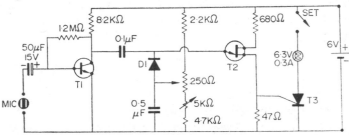

Fig 52 *The following semiconductors are used:* T1 = BC108; T2 = 2N2646; T3 = CRS1/05; D1 = 1N914. *Any near equivalent to* T1 *or* T3 *will suffice. This circuit is reproduced by permission of D. J. Bagwell*

In use, the telescope should be lined up on the Moon some time before the event is due, and the obscuring plate adjusted in position so that, with the drive on, the bright lunar image is completely obscured, but light from the area of the dark limb can reach the film. The drive must now be left on and the telescope left clamped. Next, the film is wound on or loaded (depending on the exact design of the camera) and the film advance motor started. If a disappearance is being measured, the star will already be forming a line image on the film; if a reappearance is the subject of study, the star will begin to produce a line on the film as soon as it appears. In either case the light flashing circuit must be activated from a telephone signal soon before the event occurs. It must be activated again as soon as possible afterwards. Finally, after another few seconds, stop the film advance motor and activate the light a third time. This final image will enable the relative positions on the film of the images of the star and slit to be determined. The position of disappearance (or reappearance) of the star at the moment of occultation is then set off along the film by this amount and the distance of the resulting point from the time marks whose times are known can be measured. Simple interpolation then gives the moment of occultation. Clearly this method will be more accurate if the film advances more quickly, but this in turn calls for more sensitive film as the effective exposure is less. A fast film of fairly fine grain should be used and the optimum film advance rate found by trial runs.

Atmospheric turbulence. Little need be said here. Any suitable group of objects—the Pleiades, Saturn and its rings, etc—should be photographed at intervals of somewhat less than one second on a poor night. Successive negatives may be compared with a cine projector, when the relative movements of different objects in the field due to low frequency turbulence in the atmosphere will be seen.

The Sun. The Sun is particularly easy to photograph. Many amateurs already take photographs of sunspots at irregular intervals to watch for and record developments. All that will be necessary is to make the exposures at regular intervals and to avoid, as far as possible, any relative movement of the solar image and the film between exposures.

If photographs of the complete solar disk are taken at a rate of, say, two per day around times of sunspot maximum and projected at about five to ten frames per second, both the solar rotation and the development of sunspot groups can be illustrated.

*Jovian atmosphere. The owners of larger instruments have been regularly obtaining good photographs showing detail in the atmospheric belts of Jupiter. Owing to the rapid rotation of the planet, exposures may be made at a rate of about six per hour and projected at over ten per second to show, as for the sun, both the development of features and the planetary rotation.

Phases of the Moon and Venus. The production of films to demonstrate the phases of these bodies is clearly straightforward. It may be remarked that the Venus film will aptly demonstrate the considerable change in the apparent size of the planet between inferior and superior conjunction, while the lunar film (if exposures are made for the same time with similar atmospheric conditions on each occasion) will even more vividly show the rapid increase in brightness just before, and decrease after, full Moon. The size of this effect is rarely fully appreciated.

Variable stars. Again, the original negatives must all be made under the same conditions of sky transparency as far as possible, and given the same exposure. The assembled film will show the changes in brightness of any variable stars in the field. The measurement of the magnitude of variable stars is usually done visually or with photo-electric devices, and this photographic method is unlikely to provide accurate results. However it will illustrate very well, for example, the changes in the brighter long-period irregular variables. Reference (1) gives a list of these stars.

*Motion of comets. Careful photography of a bright comet with a wide field camera would provide material for a film showing both the movement of the comet among the stars and the development of features in the tail.

Jovian satellites. The Galilean satellites of Jupiter are not difficult to photograph, and it is easy to produce a film showing their relative movements, and eclipses and transits in the system.

*   *   *   *   *

No attempt has been made in this chapter to go into very great detail about any one of the suggested projects: the main problems that must be considered have been pointed out and the general method of solving them indicated. Much will depend on the type of equipment that can be bought or made and the size of the telescope and mounting available.

There is great scope for originality and invention in work of this sort, and those who attempt to build a camera are likely to learn as much about simple engineering as they do about astronomy. This should not, however, deter any astronomer with access to simple tools from trying out at least one of the projects mentioned here.

Part 2
AMATEUR
OBSERVATORIES

Introduction

UP TO NOW WE HAVE BEEN CONSIDERING TELESCOPES OF MANY
kinds. Some are portable; others are not. And once a really
powerful telescope has been obtained, the problem of a permanent observatory has to be faced. Taking a telescope to pieces
at the end of each observing session is a wearisome business,
and also a dangerous one; it is only a question of time before
something is dropped, with disastrous results.

The length (and weight) of a refractor increases very rapidly
with aperture. A 4in can be carried, but not easily. I would regard a 5in as non-portable, so that it must be set up on a permanent mounting outdoors and taken inside only rarely (if at
all). Anything larger is unquestionably a fixture. With Newtonian reflectors, I would say that a 4in is portable, a 6in portable with difficulty, and an 8in non-portable. Cassegrain
reflectors are easier to handle because they are so much shorter,
but not many amateurs have them because they are expensive
and tricky to handle. I suppose it is true to say that over 95%
of all amateur-owned reflectors are on the Newtonian pattern.

A refractor is relatively sturdy, and unless actually dropped
or badly knocked about it is unlikely to go out of adjustment.
Not so with the reflector, which can only too easily go out of
alignment. I know of several amateurs with reflectors of
moderate size—around 8in aperture—who leave the tubes and
mounting outdoors all the time, and take out the optics, or at
least the main mirror, at the end of the observing period, so that

the most delicate optical components can be stored safely indoors. This, as I have said, is not a good idea, particularly since the dismantling process has usually to be done in the dark. Moreover, the mounting will suffer from the weather. Rust has a habit of appearing almost overnight.

One method, of course, is to cover the whole telescope with waterproof material, such as tarpaulin. This is better than nothing at all; but damp can always get in, and personally I would use this arrangement only as a last resort.

I have tried out one 'dodge' which I have not seen anywhere else, and which has not been unsuccessful on the whole. I have a 4in refractor on an equatorial head, which I use mainly for observing the Sun. For it I devised what I call the umbrella arrangement (see Plate 4). The telescope head is fixed to the top of an iron pillar, and lower down a cylinder is fixed to the side of the pillar; into this fits the handle of a large tin container which covers the telescope completely. To stop the whole thing from revolving in the wind, a hole was bored through both the cylinder and the container handle, so that a bolt could be put in. To prepare the telescope for use, simply take out the bolt and lift the container bodily off. It can be awkward to get back in the dark, and of course it is not damp-proof, as the container has no bottom; but it does help. I admit, though, that I would not use it were that particular 4in one of my main telescopes.

However, let us now turn to actual observatories. There are various forms; but before discussing them, due attention must be paid to the choice of a site. Here the amateur, unlike the professional, is frequently at a disadvantage, because he has no control over things such as street lights and neighbours' trees. Again, much depends upon his own special interests; but one principle is always to be avoided—that of the roof-top observatory.

There seems to be a widespread impression that an observatory ought to be as high up as possible. This is all very well when we are selecting a site for a major telescope; on the summit or upper slopes of a mountain, the air is thin and relatively clean, so that conditions are vastly improved. This is why so many great observatories are mountain-sited (Palomar, for instance, and Mount Wilson). But going up several thousand feet is one thing; going up ten or twenty feet is another, and in general is of no advantage at all, apart from the occasional possibility of reducing obstructions by tall trees or adjacent buildings.

Against this, an observatory perched precariously on a roof will be subject to vibration, and it will receive the full fury of whatever wind happens to be blowing. Worse still is the problem of rising hot air. If the house is inhabited, it will be warmed in the winter; the hot air from it will swirl up over the roof, and the effects upon seeing conditions will be devastating. This is also why attempts to observe by poking telescopes out of bedroom windows are in general doomed to failure.

It is vastly better to set up the observatory on terra firma, as far away from inhabited buildings as is possible. Alas, difficulties always occur. As likely as not, any tree or building will be found to lie in the most inconvenient direction possible, and if there is a street light nearby it will shine straight across that area of the garden where the horizon is most nearly free from obstructions.

In general there is very little that can be done about this except put up with it, and select the position which is thought to be the best (or, rather, the least bad). The lunar or planetary observer will need his best view to the south, and may have to sacrifice the whole of the north or east. The variable star enthusiast will ideally need a view of the whole sky, but will have to work out which site will give him the most reasonable scope. A street light can be screened, always provided that the shield is put up in one's own garden, but outside trees, together with glows caused by town lights, simply have to be borne with as much patience and fortitude as possible. Of course, it is essential to do some hard thinking before starting work on any permanent observatory. To put down a concrete base, only to discover that it would have been much better laid in another part of the garden, is not a procedure to be recommended.

The next consideration is, naturally, the form of observatory to be built, and here the type of telescope is all-important.

Those people with special needs, or with special abilities, can of course launch out into really sophisticated designs. I am thinking particularly of the heated observatory designed and used by Dr E. G. Hill, of Swanage, England. The telescope is a 24in refractor on an utterly unconventional altazimuth mounting. The telescope is, in fact, outside the observer's cabin, and the whole observatory revolves, so that the telescope itself will move only in altitude. It is absolutely fascinating, and it is extremely effective, as I know from my own experience with it and with the similar observatory used by C. A. Cross at Northwich, England (a 12in reflector, in this case). However, it is obviously difficult to make. Anyone who wants to try should first consult Dr Hill's own description of it, 'A Practical Heated Observatory', E. G. Hill, *Journal of the British Astronomical Association*, Vol 72, p 102ff, 1962.

Observers of the Sun will have to ensure that everything is light-tight, and here a dome is certainly the best answer. Also, an equatorial, clock-driven refractor is particularly suitable. This, of course, refers to the really serious solar worker who wants to take detailed photographs. More casual 'Sun-watchers' such as myself need not be so rigid in their requirements!

For a moderate-sized reflector (say 6 to 12in aperture) the run-off shed arrangement has much to recommend it. It is easy and cheap to make, and easy to use. The disadvantages are twofold. First, there is no shield against stray light—and the glare from a window, even if heavily curtained, can be most annoying. Secondly, the observer has to work in the open. On a winter's night, a gentle breeze can assume the character of an Arctic blizzard, and moreover any appreciable wind can shake even a firmly-mounted telescope. One soon becomes used to this sort of thing, but it is idle to deny that observing from inside a dome or a run-off roof arrangement is distinctly more comfortable.

Run-off observatories are described in detail by Henry Brinton on pp 216–25. Meanwhile, it is enough to note that they are of two main kinds: those in one piece, with a door or removable section at one end, and those made up of two sections which can be pushed together and latched in the middle. I prefer the latter, probably because I have used one ever since 1947 for my $12\frac{1}{2}$in reflector, and have never had the slightest trouble with it. The material may be wood, hardboard or even plastic with a firm frame; if no proper rails are available, angle-iron will serve. Originally I had a run-off shed in which there were no rails, but which had wheels moving along a trough in the concrete base. This worked reasonably, but was always liable to stick, and rails are certainly better.

An extra advantage is that a run-off shed is easier to move if one shifts from one home to another. All that has to be done is to smash the concrete and remove the rails for relaying on the new site.

There is one more limitation, however. When in its 'rest' position, a conventional reflector is low-slung, whatever be the nature of its mounting. For a moderate-sized refractor (4in or over) no run-off observatory is really suitable. The telescope, on its mounting, will be tall; anyone who sets up a refractor on a low stand will have to become an expert contortionist before he can hope to use it on any celestial object which is high in the sky. Any attempt to take the telescope off its mount every time an observing session has ended will result in frustration, bad temper, and eventual damage to the instrument. A run-off shed for a refractor of this type would have to be tall and slender, thereby being ideally suited to catch every scrap of wind-force. It would also be awkward to move. Of course, run-offs for refractors cannot be ruled out, and they do exist; but when I considered what sort of housing to give my 5in refractor I decided in favour of the run-off roof arrangement.

Here, the opposite applies. Because the refractor is high-mounted, the walls—which do not move—will not cut off the horizon to any marked extent. A little may well be lost, but by the time an object has dropped below 10° or so there is not much point in trying to look at it in any case. On the other hand, the permanent walls will be crippling from the point of view of the conventional reflector. Unless the telescope is mounted on a tall pillar, so that the observer has to mountaineer in order to reach the top of the tube, there will be no sky available below around 20° to 30°, which will be wildly inconvenient. (Bear in mind that I am referring to a Newtonian reflector, not to a Cassegrain or a Gregorian.)

Run-off roof observatories are described by F. R. Spry on pp 226–35. Again there are various types. Roofs which are wholly removable are remarkably awkward to handle, and the roof material has to be lightweight, which means that it is not likely to be waterproof. Hinged roofs have marked drawbacks. No section, on a hinge, can be really large, and the whole arrangement comes under the heading of 'bits and pieces', which is never really workable. A roof which slides back on rails is far better; it may be in one piece or two. On the whole it is fair to say that anything which flaps is best avoided.

So far as materials are concerned, the observatory may be of wood, hardboard, or plastic with a firm frame. The roof may be rolled back by a pulley arrangement; observers with real practical skill may prefer to motorise it. If well made, the whole construction can be fully weatherproof. Beware of allowing snow to accumulate on the roof in winter; a few judicious strokes with a long-handled brush may save a great deal of trouble later on!

The ideal kind of observatory is, of course, the dome, and the various types are described below by J. Hedley Robinson. It is fair to say that the actual building is not really difficult for anyone gifted with his (or her) hands, but there is one major hazard: the circular rail. As is explained in Chapter 16, not all domes run round on rails, but my own does so, and it works admirably. The rail is made of aluminium strip, and was bent into the correct shape by using an ordinary vice with three nails. By fixing two nails, one on each end of one face of the jaws of the vice, and the third nail in the opposite face, it was only necessary to slip the aluminium strip through, an inch at a time, with exactly the same pressure on the vice handle. The revolving platform on which the dome stands is made of $1\frac{1}{2}$in angle-iron. The structure is octagonal, and the eight frames are bolted together. The roof also is made of eight sections, one of which opens and is made in two parts. To open, there is a long metal rod attached to the opening side of the hatch, and there is a wooden block fixed to the hatch close to the hinge side. The hatch is lifted to a vertical position by the rod, and the rod is then rested against the wooden block, so that the hatch can be lowered gently on to the roof. The lower hatch can be opened by hand, from the outside. Another refinement is a brake unit on the inner circular rail when 'on', two blocks grip the rail and prevent the whole dome from spinning round in the breeze. The material used for the main observatory is wood; the vertical sides of the 'dome' are of glass held in light frames. I wondered whether any trouble would be experienced from daytime heating, but I have found none, and certainly the glass makes the whole building somewhat decorative—as the photographs show (Plates 12 and 13). Let me add that I did not build the observatory myself. It was done by R. A. G. Gulley and B. Gulley, my cousins, and has been described more fully elsewhere. ('An Observatory Dome for Amateurs', R. A. G. Gulley. *1971 Yearbook of Astronomy*, Sidgwick & Jackson, London 1970, p 111ff)

When a dome has been built, not much maintenance is needed, except for obvious precautions such as keeping the moving parts well oiled. Things are even simpler with the type of observatory in which the entire structure revolves, and the circular track lies on the ground. These too are described below, but I admit that personally I do not like them. Theoretically they are excellent, but in practice they always seem to jam solid at inconvenient moments, and of course there is much more weight to be revolved.

There is one more point to be borne in mind. An observatory which simply stands, unattached, on a concrete base (as my own does) is not a 'permanent building', and on this score no planning permission for it is required. Local officials who make themselves awkward about 'building lines' and restrictions can usually be chased away easily. And if the owner decides to move house, he can take his observatory with him. But once there is anything anchored down, the situation is much more difficult. Planning permission must be obtained; and if you sell your home—beware! I had a relevant experience myself in 1968, when I moved from Armagh, in Northern Ireland, to my present permanent house in Selsey. I sold the Armagh house to a local veterinary surgeon. When the deal had been completed, he calmly informed me that since my dome stood on what had now become his property, he proposed to claim it; I gather that his wife thought it would make a nice summerhouse. Of course I dismantled it immediately and shipped it off to Sussex. In fact I was on the right side of the law, because there was no fixing to the ground; had the sides been concreted in, I would have had to take illegal action—which I would most certainly have done! The golden rule is: Before selling your house, either dismantle the observatory before signing the contract, or else have a written clause in the agreement. Reluctantly, it must be said that nowadays a gentleman's agreement is not sufficient.

There seems little point in saying much about the many refinements which can be added to an observatory, whether it be a run-off roof arrangement or a dome. It is always wise to have a source of light to hand; it need not be brilliant—in fact, a bright light is harmful, as it destroys one's night vision for some time—and a torch-bulb, powered by an ordinary torch battery, is sufficient. (Where the telescope is on an electrically-driven mounting, there will of course be a power supply laid out to the observatory, and any desired lighting can be arranged.) Shelves are always valuable, even in a run-off shed. Observing steps are also very desirable. My own are shown in the photograph (Plate 9), but are admittedly rather cumbersome, and an adjustable chair is often recommended. When one is observing for long periods, it is only sensible to make oneself as comfortable as possible.

In short, then, it may be true to say that:

1. A very small telescope (below 3in aperture for a refractor or 6in for a reflector) needs no observatory. It can be kept indoors, and taken out when needed.
2. A moderate reflector can well be housed in a run-off shed (see Chapter 2). This is not so suitable for a refractor.
3. A moderate refractor is well suited to a run-off roof arrangement (see Chapter 3). This is less suitable for a reflector, though of course it can be adapted to any telescope.
4. A dome is much the best form of observatory (see Chapter 4), but is of course more expensive, more difficult to make, and is essentially permanent when set up.

14 : HENRY BRINTON

Run-off Observatories

FOR PEOPLE WHO ARE FORTUNATE ENOUGH TO HAVE A FAIR-SIZED telescope, and to whom money is no object, or are born handymen, the dome is the ideal form of protection. The construction, however, tends to be both expensive and time-consuming. The usual alternative is to have a run-off shed, which has many of the advantages of a dome with only a few drawbacks. It fulfils, moreover, the main qualifications required: ease of construction, low cost, and speed in preparing the telescope for use.

The essence of the run-off shed is a complete housing, which covers the instrument when not in use, and pushes off on rails when the time comes for observing. Such a construction is easily made so that the telescope is as well protected from the weather as it would be if housed indoors, but can be wholly exposed in less than a minute. In a way, one might describe the arrangement by saying that instead of keeping the instrument indoors and moving it out when requires, one keeps it indoors and moves the house!

The run-off shed can take either of two forms: it may be a single construction with one end which is a 'door' which lifts off, or it may be built in two sections joined together when the telescope is not in use, but each half of which can be pushed away in opposite directions for viewing purposes. Each method has its advantages and its drawbacks. Photographs are given (Plates 8 and 9) of my own shed, which is of the first type, and Patrick Moore's, which is of the second. Each covers a 12½in reflector.

The actual building is similar in both cases. The shed consists

216

of a wooden structure covered with weatherboarding or hard-board, with a sloping roof to enable the rain to drain off. Incidentally, builders should not fall into the error which I did when I made my own. The low side of the pitched roof ought to be on the side from which the prevailing wind blows. The force of a gale can be devastating. I have the high side on that of the prevailing wind; and though my observatory is sited in a sunken garden, with hedges all round, the instrument on its concrete base has been shifted bodily, even though the telescope and mounting weigh the best part of a ton! Moreover, in addition to this precaution, it is as well to have a properly anchored wire which can be clipped on to the shed when it is closed.

The drawback to the single construction is that one end—the 'door'—must be removable, so that the shed can be pushed back. This can be a clumsy operation with a large telescope; it is not easy to design a fastening which is both secure and easy to remove, and the shed itself will lack some structural strength as the result of being three-sided. On the other hand, and provided that the door is effective and easy to move, there is only one set of rails to make and one lot of pushing to do. Hinged doors are not to be recommended, for reasons which will be given later.

Much of what follows comes under the heading of sheer carpentry, but may be found useful. In constructing the framework of the shed there is no golden rule, except that it should be strong enough to stand up to heavy gales. Each telescope has its own, often odd, shape. Therefore it is impossible to suggest actual dimensions. Generally speaking, 2 × 2in wood is quite suitable for a medium-sized telescope such as a 12in reflector. If hardboard is to be used for the outer covering, the upright members should be designed to be in such a position that the joins will come over the framework. With weatherboarding there is no such limitation. Moreover, weatherboards have the advantage that they have less tendency to warp.

Hardboard is suitable in many ways. It is cheap to buy, and it will stand up to the weather well. My observatory is within a hundred yards of the open sea, with nothing but a hedge to keep off the winds (or, more usually, gales) which come straight from South America, laden with corrosive brine and abrasive sand. With a single outside coat of bitumenous aluminium paint, the board shows no signs whatsoever of yielding to these challenging conditions. Another advantage is that each wall of the shed is semi-rigid in itself, by the mere process of cutting it out to the right size. If weatherboarding is used, rigidity is ensured only by accurate and sturdy fastening.

For a 12in or larger telescope, the board should be fixed on 2 × 2in sawn joinery softwood along all the edges, crossed if need be with 2 × 1in for extra rigidity. For smaller telescopes, 2 × 1in may be strong enough throughout. It is as well to creosote the wood after cutting, but before joining and nailing or screwing on. As the wood is liable to give first at the joins, treating at this stage will protect the most vital parts.

Because the wood is there principally to give extra rigidity and to join the sides, there is no need for neat joints. The board, being nailed or screwed on to the wood, will keep the battens in place. A simple joint for joining the top and end supports can be made by cutting out half the thickness of the batten at each end, and then nailing one over the other to fit flush, Fig 53 (a).

Hardboard comes in sheets of 12 × 4ft, and it is therefore necessary to have joints. Needless to say, it is most important to fix a batten behind each join; but, for the roof particularly, just nailing the board on to the wood is not sufficient. Before putting the batten in place for nailing, it should be laid with a strip of sealing compound to come immediately under the join. Selastic, or almost any proprietory seals, will make a good, waterproof join if applied in this way. Otherwise, it is miraculous how the water will find its way through.

The run-off shed consists essentially of three fixed sides, a roof and a door (assuming that we are to make the first type, not the two-piece). The simplest way of going about the construction is to make and erect the three fixed sides first. Each side will

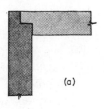

(a)

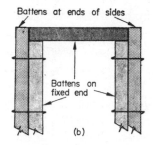

Battens at ends of sides

Battens on fixed end

(b)

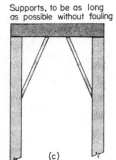

Supports, to be as long as possible without fouling

(c)

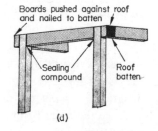

Boards pushed against roof and nailed to batten

Sealing compound

Roof batten

(d)

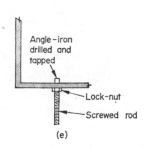

Angle-iron drilled and tapped

Lock-nut

Screwed rod

(e)

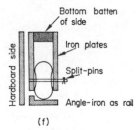

Bottom batten of side

Iron plates

Split-pins

Hardboard side

Angle-iron as rail

(f)

Fig 53 *Constructing the framework of the shed.*
See text for fuller explanation.

consist of a sheet of hardboard, fixed on battens, as described. One side is, of course, lower than the other, to allow for the pitch of the roof (15° is quite sufficient), and the fixed end has to be tailored to fit the two sides. The actual design and method of fastening will have to be decided in the light of the dimensions of the individual telescope.

The fixed end can be fastened on to the sides in a number of different ways. One is to make the battens of the fixed end fit snugly inside the battens of the two sides (b), and then bolt or nail the two sets together. This, though reasonably good, will not be quite firm enough for larger sheds. For these, small diagonal pieces can be fixed across the corners at the top and bottom, their size being dependent upon the amount of clearance in any individual case.

The open end, where the door is to be hung, presents a more difficult problem, since by its very nature it can have no support whatsoever across the bottom. If it were not that the rails prevent the sides from splaying outward, a run-off for any but a small telescope would be impracticable in this form. As things are, it is necessary to design the whole shed so as to allow sufficient clearance to permit small diagonal supports across the corners from the sides to the roof (c).

The question of the roof needs careful consideration, since it must be absolutely waterproof both in itself and where it joins the walls. It is not likely that a single sheet of whatever material is used will be large enough for the purpose. Hardboard tends to sag and allow pools of rainwater to collect, and the join, even when there is a beam of the frame to which the edges can be fastened, tends to leak. If this method is used, the edges of the sheets should be smeared with jointing-compound before being attached to the beam, and it is a good idea to have a strip of felting compound glued over the join. Indeed, it is wise to do this with all the joins, to prevent water from seeping in and rotting the wood of the frame.

The board can be fastened straight to the battens running

along the tops of the three fixed sides, with its own battens cut to fit closely inside these (d). A couple of inches at least of overlap of the roof board should be left all round. When it has been fixed, flat pieces of board, 3×1in or $2\frac{1}{2} \times \frac{3}{4}$in, should be pushed up tight against the overlap of the roof and nailed on to the battens of the sides with sealing compound in the joint.

An alternative material for the roof is serrated glass fibre. This has an added advantage of giving more light when one is inside the shed, but the material has an unpleasant habit of cracking if roughly treated.

The door is largely a matter of choice. There is little or nothing to be said in favour of hingeing it. Whether or not it is divided, a hinged door will only flap and swing and get in the way when the shed is pushed back. Much the best plan is a door which will simply lift off, though, if it is big, it can be a nuisance in a wind, acting as a powerful sail which will suddenly carry off the observer if he is unwary. On the other hand, few of us have telescopes rigid enough to be of much use in a high wind, so that this particular hazard is not really likely to arise!

A door which lifts off can be made like the lid of a box. The top and side battens of the shed can be flush with the board, and the door can have battens which fit outside these, like the rim of a box lid. Better, I think, is to let the boards of the shed stick out beyond the battens, and let the door fit inside, with its own battens lying up against those of the shed. Any form of fastening can then be made, either round the outside, or through the two pairs of battens.

So much for a shed. Next we must consider the rails, which, though essentially simple in principle, are perhaps the most time-consuming part of the entire building operation.

Builders' merchants can sometimes produce rails intended for sliding garage doors; but these are not easy to find, because most garage doors nowadays either hinge, hang or roll. In any case, these rails are expensive. A simple solution is to use ordinary $1\frac{1}{2}$in or $1\frac{1}{4}$in angle-iron, which is cheap and satisfactory.

Naturally the rails need fixing, and the strength and closeness of the fasteners depends upon the nature of the ground. The easiest method, if you have a friendly blacksmith close at hand, is to get him to weld a number of rods or strips along the bottoms of the angles. The perfect way of laying is to sink each of these in a concrete pocket, in which case spiral strips should be used. If the ground is firm clay, it is enough to have 6in rods pushed into the clay. Most builders' merchants can now provide ready-mixed cement in small bags, so that laying the cement pockets is an easy matter. In either case, it is essential to make sure, by using a gauge, that the rails are truly parallel and the right distance apart.

If there is no blacksmith nearby, simple fasteners can be made with threaded steel rods, easily obtainable. This involves drilling and tapping a number of holes along the rails. When this is done, the rods must be screwed to the underside of the rails, so that the tops of the rods are flush with the surface of the rails (e). Lock-nuts must then be pulled up tight on the underside.

Obviously, the boards of the sides are continued down beyond the 2 × 2in battens which carry the wheels. Hardboard is quite rigid enough for this; but slots will have to be cut to allow the axles to protrude slightly. Also, it is wise to screw or bolt on both sides a single piece of iron strip to the 2 × 2in on the outside, and bend it inward slightly, so that it will just rub the edge of the rail. In this way, the wheels prevent the unsupported front ends from splaying outward, and the strips from splaying inward.

Wheels are fairly easy to come by. They may be of some hardened rubber compound, say 5in in diameter with 1in axles, meant for fitting on trolleys. They can be bought at most large ironmongers.

A simple way of fitting the wheels on to the bottom of the sides of the shed is to have pairs of plates cut and drilled by a blacksmith. Through two holes in the top, the plates can be bolted on either side of the 2 × 2in running along the bottoms of the sides of the shed (f). One-inch holes, drilled opposite one another in the bottoms of the plates, can take a simple piece of steel bar to form axles for the wheels. The axles can themselves be held in place by drilling small holes through the ends of the bars and inserting split-pins when the wheels are in place.

There is not a lot to say about the two-piece shed. There is the disadvantage that two sets of rails have to be laid (or else one much larger set). There is also the need for making the join between the two halves watertight. This is achieved by having an overlap on one half, which will fit snugly over the other half at the join. If it is too tightly made, warping may prevent the shed from closing. If it is too loosely made, rain may get in at the gap. However, with proper care, the two-piece shed has distinct advantages, the chief of which are having no door to deal with, and the greater strength, for size, of the separated halves.

With either type, remember the strength of the wind. If the shed is narrow in relation to its height and length, there may be a danger of its blowing over in a high side-wind. There are various ways of dealing with this hazard. One is to sink a metal rod, with a loop or ring on top, into the ground on each side. If hooks are then fastened to the sides of the shed, they can be so arranged that they will slide into the rings automatically, and so anchor the shed without the need for any action. In any case, any fastening should be quick to release. There are enough things to fiddle with in preparing the telescope without wantonly adding to the number.

One of the disadvantages of the run-off shed as against the dome is that with the latter, fittings can be incorporated to hold all the impedimenta used in observing—charts, filters, spare eyepieces and so on, all of which are close at hand when needed. This lack in the run-off shed can be compensated for with a little ingenuity. Edged shelves can be fitted at the near end of the run-off portion or portions, which will still be within reasonable distance of the observer. The great temptation to resist is that of putting down some delicate piece of equipment, such as a filar

micrometer, on the roof of the shed, and then forgetting to remove it before the shed is run back over the telescope! For the
really absent-minded—and astronomers are such by tradition—
it is worth having a folding table which folds into the shed wall
and comes out on an extended arm when one is working at the
telescope.

One or two words of caution may be added if a run-off shed
is to be used. One is that in hot weather the inside can get very
hot indeed, even when some ventilating system is used, and it is
desirable to run the shed off, if possible, at least an hour before
the telescope is to be used. Otherwise thermal currents of air,
due to the heated ground and mounting, will ruin observing.
With a dome the same thing can happen; but the amount of
temperature change is generally much less where all that is done
is to open what is in effect a window.

Another point to be watched is at the design stage. The rails
upon which the shed is to be run must be long enough to take
the shed well away from the telescope when it is run off. Few
things are more exasperating than to find that some object of
great interest is out of view behind the roof of the shed. It is one
of the laws of nature that clouds will come up just as the object
emerges!

When designing the shed, it is not absolutely necessary that
the line of rails should be parallel to the axis on which the telescope is normally rested when not in use. Either the barrel of the
telescope can be left at right angles to the mounting, or else the
shed can be made wide enough to house the length of the tube in
the breadth of the shed. In fact, one should consider all the
possible problems, such as prevailing wind direction, obstructions to viewing, and available space in different directions
before deciding upon the form and dimensions of the shed.

Every individual telescope and observatory site will present
its own peculiar problems, but a very little forethought and
planning will be able to cope with them. If such care is taken,
the run-off shed is a thoroughly satisfactory method of housing.

It need not daunt the novice to contemplate erecting one for himself. If I, being one of the least handy-men living, was able to make a run-off shed which has lasted happily for fifteen years so far, there can be very few indeed who could not do at least as well.

Run-off
Roof Observatories

THE RUN-OFF OBSERVATORY DESCRIBED ABOVE IS EXCELLENT FOR a reflector. For a moderate-sized refractor, however (say a 5in, or even a 4in) it is frankly unsuitable. Unless a dome is to be built, the best answer is therefore a permanent observatory with a run-off roof. This, in its turn, is not well suited to a Newtonian reflector of the 6 to 12in range. The walls must be fairly high; and unless the reflector is mounted on a tall pillar, so that the observer has to mountaineer before reaching the eyepiece at the top of the tube, there will be no sky available below 20° or 30°.

In short, the run-off roof arrangement is best adapted to the needs of a refractor, though of course it can be modified so as to cope with a reflector as well.

Again there are various patterns of run-off roof observations. One involves the roof being made to slide back in two halves in opposite directions. But for the moment it may be best if I describe the observatory which I have myself constructed to house a 5in refractor. As will be seen from what follows, it does not present too many difficulties to the average handyman.

The telescope was set on a pillar with equatorial head, and with the telescope at rest in the horizontal position the height was 6ft 3in. The tube, complete with dewcap, was 5ft 6in long, and this meant that the eyepiece would have a varying height range of approximately 30in from the horizontal to the near-vertical.

As the observer (also the Editor of this book!) happens to be

6ft 3in tall, the height of the mounting had to be arranged to suit him. Raising the stand about one foot, on a massive concrete block, gave comfortable viewing for a tall observer with the telescope at 45°, and this decided the basic height for the sides of the building. Bearing in mind the frequently strong prevailing south-west winds, the building was made as low as possible without sacrificing efficiency.

Having decided the approximate height of the sides (8ft), what size should the base be? It is good to have plenty of room in which to operate, but the larger the floor the larger the roof. Theatre audiences are often invited to 'Raise the roof!' but my problem was to have it disappear altogether (and subsequently reappear) at short notice, and it was clearly best to keep the roof as small as possible without restricting the view and working space. That brought me to the biggest problem: What type of roof?

Various ideas were given careful consideration. Should it be in one piece, or two, or more? Should it hinge, fold or slide? With due regard to conditions of wind and weather, and bearing in mind that the operation would have to be easy (and almost always carried out in the dark), some patterns were rejected out of hand. Hingeing certainly meant at least two pieces, if not more, and each piece would start to operate somewhere about 8ft from the ground; this would mean special fastenings to stop them from flapping in the breeze when down. Folding would need a great many hinges, together with the problem of making the whole construction weather-proof.

That left a roof which would slide. It could be in one piece, giving little wind resistance; it could be secure on its rails all the time, and when rolled back it would expose the entire top view with no obstruction at all.

With the refractor at horizontal rest, 8ft square gave sufficient room, and more of course if it were elevated; so the decision was made to make the building 8ft square, for both sides and roof.

A concrete base 8ft square, with a central portion 3ft 6in square raised by 1ft for the refractor to stand on, was the first operation. The construction of the sides can depend on personal inclination and ability, together with material available and the cost. Bricks and mortar naturally give the most substantial result, if one is capable of this work, but a well-designed timber framework, suitably covered, should give all the protection necessary. Hardboard, corrugated plastic sheeting, and wood such as weatherboard are all suitable, and are easy to cope with. With some help from a local carpenter, the sides were constructed of 3 × 2in timber to the design shown in Fig 54. This, of course, can be varied so long as good diagonal bracing is incorporated to ensure stability—and do not forget to provide for a door! (It is surprising how easily this can be overlooked. Also, do not make the door too narrow. Bulky equipment may have to come in and out.)

The roof was constructed of the same 3 × 2in timber to the design shown in Fig 55. This also may be varied according to personal inclination.

Because the site was well sheltered (there is a convenient hedge to act as a wind-shield in the most vulnerable direction), light plastic sheeting was chosen for both the sides and the roof. Two lengths of angle-iron were fixed to the sides of the building. These were 17ft long, projecting to the north side on suitable supports, not forgetting the need to have sufficient slope to allow rainwater to drain off and avoid having a showerbath when starting to roll the roof back. The overall impression is shown in the photograph (Plate 10).

Three roller bearings were fitted to each side of the roof in suitable housings, extended to curl under the angle-iron rails to guard against possible wind-lift.

When at last all was in position, there came the problem of opening the observatory. Two wooden winders, with cord running over small pulleys, were tried as an experiment. They worked after a fashion, and by fitting a roller bearing on the

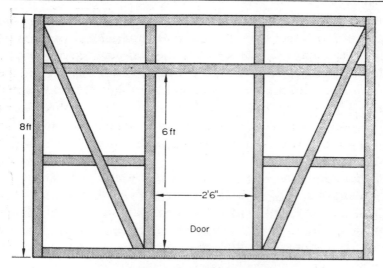

Fig 54 *Construction of the sides*

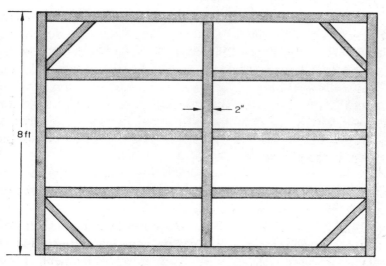

Fig 55 *Construction of the roof*

centre of the north face to support the centre roof member (a strip of 1 × 1/16in steel screwed to the underside first) the running was improved, because this cancelled out the slight sag of the 8ft span. All the same, it was not really satisfactory.

We next tried using a small winch with flexible wire running over larger pulleys, after the style of window or theatre curtains. This was an improvement, but a central point of push-pull over such a large span was not really a good idea. An even pull on both sides at the same time was needed. To do this by the same method would entail wire and pulleys all over the sides and roof, and, knowing only too well how everything will catch in everything else when not required to (especially in the dark), some radical re-thinking was needed. The essential thing was to have something more substantial, as foolproof as possible, operating on both sides at once, and not too complicated.

Cycle-chain running around suitable sprockets connected by a shaft seemed likely to give most of the right answers. Material, cost, and the necessary machine-work involved was the next consideration. We came to the conclusion that we needed:

Six cycle hub sprockets (20 teeth).

About 40ft of cycle-chain.

8ft of shaft ($\frac{5}{8}$in diameter mild steel was deemed stout enough).

A piece of 1$\frac{1}{2}$in diameter mild steel for making bushes.

Bearing for shaft.

Odd pieces of angle-iron and rod.

A crank handle, and a housing for it.

Various nuts and bolts, etc.

The estimated cost was about £10, which proved to be remarkably accurate. It must be added, however, that I do have a well-equipped private workshop, including a suitable lathe (a Myford ML7), and I could do the necessary machining without having to call in outside help. This, obviously, made things a great deal easier than would otherwise have been the case.

The sprockets, designed to screw on to back-wheel cycle hubs 1$\frac{3}{8}$in diameter, 24 threads per inch, needed bushes to adapt them to the ⅝in shaft, and these were first on the list. The 1½in bar was reduced to 1⅜in, just sufficient to take one sprocket (Fig 56), and threaded 24 TPI to screw on fairly tight, leaving a shoulder of ⅛in. It was then reduced to 1⅛in for another three-quarters of an inch to take securing grub screws, drilled and bored ⅝in to fit the shaft and parted off. Another similar bush was then prepared, with enough thread to take two sprockets. Next came two more bushes for the sprockets intended to keep the chain extended, with must enough thread and a shoulder upon which to home, drilled 5/16in for a bolt to pass through as a stub axle.

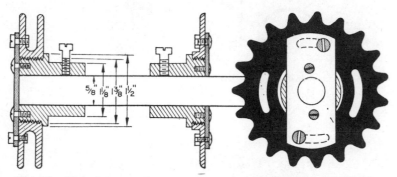

Fig 56   *Sprockets used in the device for opening the observatory*

Mounting these extension sockets will vary according to the structure of the building. I used a short length of 1in angle, 5in long, drilled and countersunk on the inside as near the web as possible for 2BA screws to hold a 2in piece of ½in square bar fixed flush with the outside angle—not central, but toward the end of each angle, ie right and left hand. The square bar was drilled 5/16in BSF to take the bolt passing through the bush forming a stub axle. Various holes were drilled in the angle for wood screws to secure to the underside of the top timber. These were fixed right in the corner, giving minimum clearance of the teeth for the roof to have maximum travel. When these sprockets were screwed on the bushes as tightly as possible, they were centre-popped all round to prevent them from unscrewing.

This was found to be sufficient, as there can be no strain on them; they only serve to keep the chain in position (Fig 57).

One more bush was required for the crank-handle assembly. This was similar to the first: one sprocket, with room for set screws, and again bored ⅜in.

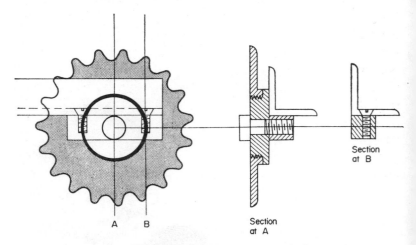

Fig 57 *Mounting the extension sprockets*

A discarded 'Picador' light alloy grinding/drilling spindle was found in my 'odds-and-ends' box, and was put into service. The last bush was mounted on a short piece of ⅜in rod. A left-hand cycle pedal crank fitted nicely on the short shaft; the pedal spindle-hole was sawn off, and a handle fitted. A file handle was handy (just the right size and shape); it was drilled through for a ¼in bolt, which proved very satisfactory. This was mounted at a convenient height just inside the door, with a length of chain to the inner sprocket of the double set on the shaft.

Bearings for the shaft could easily be made with just a piece of angle drilled through, but light alloy bearings of the 'Picador' type are fairly inexpensive and easy to fix. I preferred to use these, as they have oil retaining bushes and lead to a more satisfactory result.

Thus almost all the fittings were ready except for attaching the travelling chain to the roof. I wanted the roof to have the fullest possible movement, and it took some time to work out how to connect it in a manner which would please me. Of course, any kind of spike reaching into the links of the chain would work, but would be somewhat unprofessional! A piece of chain lying on the bench was pushed aside while I was thinking about the problem; in so doing one link folded on itself, and the resulting triangle gave me the clue I had been seeking. A drawbolt connected to the tip of it would allow it to travel to a point above the axis of each sprocket, and I could not ask for more than that.

Forming the triangle took some thinking out, but trial and patience produced the required result. Rivet *B* (see Fig 58) was punched out of the cranked half-link *A*. (Sometimes the rivets are extra hard, and it was found advisable to grind off the rivet

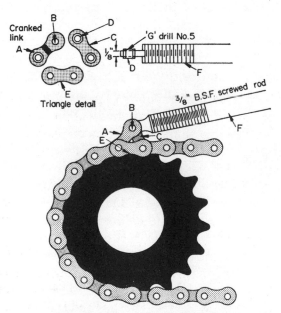

Fig 58 *Forming the triangle*

head to make removal easier and to avoid damage.) The inner link C has a hollow rivet D, which is also removed—using a suitable punch—and carefully retained. The outer roller is not required.

The rivet-hole B was No 26 drill size (0·147in) and the hollow rivet hole D was No 5 drill size (0·205in), so the draw rod F—a piece of three-eights screwed rod (studding)—was shaped at one end to fit into the link at the top of the triangle, and drilled No 5 at G to receive the hollow rivet. Links A and C were then rejoined, with the end of the drawrod between, and extra capping link plates at E. The rivets were replaced with 4 BA nuts and bolts. The pieces of screwed rod were used so that when passed through a piece of plate attached to the roof with a nut on each side, adjustments could be made to give an even tension on the chains. Two of the sprockets had radial slots in them, and these were reserved to go each end of the shaft, because it was essential that they should not unscrew when taking reverse strain. This was done by two small pieces of plate screwed into the ends of the bushes, and short bolts passing through the slots as in Fig 56.

At last all was ready for the test. The crank was slowly turned, and the roof glided beautifully along for a few inches, and then locked solid. It wound back quite easily, and was tried again, with the same depressing result. No apparent reason could be found, but when I felt behind one triangle all was revealed. One of the 4 BA bolts was a little too long, so that it was catching on one of the timber uprights of the side. It was removed, shortened, and replaced, and all was well. The roof glided out of sight without a murmur, and returned as planned.

This, as I have said, is only one of many possible designs for a run-off roof observatory. The sliding-off arrangement could be by means of pushing or pulling from the outside of the house itself, for instance; this would be distinctly clumsy, but if well made it would be workable. For a reflector, it would also be possible to make the whole top section of the building roll back on rails, though this would mean a hinged or removable section at one end. If one wants to be really sophisticated, the rolling-off could be controlled by a motor. However, the observatory described here has now been well tested, and has stood up to all its tests, so that the description I have given may well be of help to other observers who are anxious to make some sort of convenient, useful housing for a telescope of moderate size.

Observatory Domes

RUN-OFF SHEDS AND RUN-OFF ROOF OBSERVATORIES ARE RELA-tively easy to make, and within their limitations they work very well. However, there can be little doubt that a dome is by far the best form of observatory. For one thing, there is a great differ-ence between observing in the open, buffeted by every wind that blows, shaking the instrument and chilling the observer, and work carried out from a satisfactory semi-enclosed observatory.

Moreover, the dome will suit any form of telescope—it is true that a Coudé form of mounting is something of an exception, and may be regarded as the ultimate in comfort for the observer, but it is not easy to construct, and it has its drawbacks from a practical point of view. The more convenient dome is the usual solution of the problem.

Normally, the dome observatory is rather expensive in pro-portion to the use that the amateur makes of it, but if he is willing to do much of the work involved in its construction, he can produce a dome at something like a quarter of the cost of purchasing it commercially.

Domes come in two main types: those in which the whole observatory building revolves on wheels, rollers, etc; and those in which the upper portion only revolves.

The wholly-revolving type must of necessity be made of light materials, so that it can be moved easily; yet it must be strong enough to withstand wind and weather. If it is on the heavy side, considerable muscular power is needed to move the dome round during observing sessions. Internal wire bracing of the several panels in a hexagonal or octagonal structure can save weight as compared with the more normal timber construction, while the outer cladding can be of plywood or hardboard fastened to the rectangular frames. It must, of course, be made so that the running wheels or rollers do not tend to come off the circular rail or other running track when the observatory is rotated, and the author surmounted this difficulty by using rubber wheels some four inches in diameter running in a concrete gutter of semi-circular cross-section. Drain holes were provided to allow rain to run away, and the channel was deep and wide enough to allow considerable tolerance in the running of the wheels within the gutter track. Thus the wheels could climb the sides of the track for a few inches when pressure was applied to rotate the observatory, but once this ceased the wheels automatically took up their normal positions in the bottom of the gutter track. In this way, derailment of the wheels was avoided, and the construction was simplified.

The lower part of the structure was of light weatherboarding of 2 × 1in frames, as shown in the photograph (Plate 14, top). These were fastened together at the bottom of each frame by corner plates, with additional plates near the top of each frame. The wheels were mounted on plates which were screwed to the underside of each bottom member to provide additional strength to the structure at the joints of each bottom member.

The top portion of one of the eight panels was provided with an opening shutter, to enable the telescope to be used at low altitudes, while the opposite panel was provided with a door for access. One amusing result of this arrangement was that after rotating the observatory in total darkness there arose the problem of finding the door to get out of the observatory. Also, one never really knew in what direction egress would be achieved, until one became used to checking up by noting the north-south line of the telescope mounting. A garden lily-pond next to the observatory can provide yet another occupational hazard!

The roof can easily be constructed from lengths of 2 × 1in

batten, bolted to the tops of the rectangular frames, and suitably joined together to form the apex of the pyramid. The frame so produced can be covered with chicken-wire netting, with a layer of tarred roofing felt over it.

A point to be remembered in all methods of construction is that the width of the opening, however made, must be wide enough to make observation easy. Many a dome has been spoiled by having the opening too small. For normal amateur-owned telescopes, a width of about 4ft is adequate. A good portion of the sky is available, and when the instrument is reversed east to west a large sweep becomes available without having to move the dome.

If the top member of the roof construction is 4ft (plus a few inches for overlap of the joints) in length, then the opening can be rectangular and 4ft wide, as shown in Plate 14, top. The opening should be closed by two hinged sections 2ft wide each, by the length of the contiguous frames plus a few inches in length; this will provide ventilation at the top. The hinged shutters can easily be made of timber frames upon which are nailed marine plywood sheets, like a shallow box without a lid, inverted and extending upward beyond the top cross member. The construction is also shown in the photograph (Plate 14). It is wise to provide this or similar means of ventilation at the tops of the hinged sections when in the closed position, so that a through current of air can be maintained to keep the temperature of the observatory from rising too much while in sunlight. The air intake is provided by the clearance between the bottoms of the panels and the concrete ring track. The method was, in practice, found to be very efficient.

The author had a dome of this type in use for many years, but ultimately the uneven wear on the wheels, and general slackening of the construction due to constant use, brought about a desire for something more durable and easier to rotate.

After some considerable thought it was decided to make a marine plywood framed dome with hardboard cladding, and to paint the dome itself aluminium (see Plate 14, bottom). In this design, rotation is effected by arranging the dome to travel on some eighty odd golf balls—which can usually be obtained by talking nicely to the local golf-club officials!

The dome was made to an overall diameter of 11ft. The opening is 4ft 6in wide, and is covered by a flap section hinged at the bottom, and an up-and-over rolling section from the top of the flap to a point just beyond the zenith (Plate 16, top). This section runs on old motor-car ball races obtained from the local garage. These measurements are adequate to accommodate a 10in reflecting telescope, with a minimum of dome movement during normal observing sessions. The construction and design of this particular dome was undertaken by Mr Charles Merrilees, of Dawlish in Devon, at a total cost of about £50. The design itself can be followed from Fig 59.

The original octagonal walls of the rotating observatory described above were cut down to a height of 4ft 3in, and a circular wall plate was mounted on top of the remaining octagon. This wall plate is made of $\frac{1}{2}$in marine plywood, with all joints staggered and supported by further marine ply sheets underneath. The great point to remember is that this wall plate must be accurately levelled all round, because on its upper surface the balls supporting the dome must travel easily.

Of course the height of the octagon walls can be arranged to suit the mounting of the telescope, but sufficient headroom should be provided, so that the observer does not run the risk of walking into the dome frames in the dark.

The wall plate must be adequately supported to enable it to carry the weight of the dome without distortion. This can be achieved by providing supports alternately inside and outside the octagon panels, and near the centre line of the wall plate.

A second circular plate, 4in larger in diameter than the wall plate, is next provided, to form the bottom ring of the dome itself. Golf-balls, or some similar arrangement between the two plates, form a giant ball race, which revolves easily and is the base of the dome framework. A fillet of ½in high batten all round outside the ball track keeps the balls from escaping outwards. This fillet is fixed on the underside of the dome base ring. A similar fillet runs round inside the balls on the wall plate, to stop the balls from escaping inwards. Sufficient clearance should be provided to allow the balls to run freely without overtaking each other (see **Fig 59**).

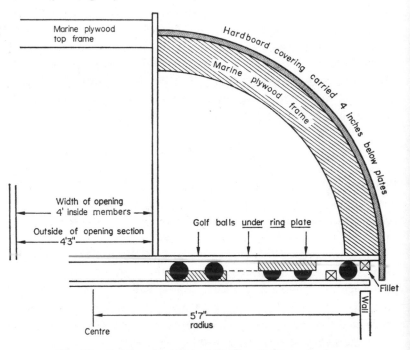

Fig 59 *Section drawing of Hedley Robinson's dome*

The dome frame members are mounted on the upper ring plate, by means of metal brackets. These all come to meet the two vertical half-circle members forming the sides of the opening, and serve to hold these members rigidly vertical to the ring plate. The up-and-over section runs by means of the ball races on top of these two semi-circular members.

The hardboard cladding is cut and fitted to the frames after they have been erected. It is interesting to note that all the sides of the cladding sheets are curved. This arises through the frames being of circular section and the cladding forming chords between them.

The number of frames provided is optional. The more frames included, the more circular the dome becomes; but it is also true that the more frames provided, the more joins are needed in the cladding. The joins between the cladding sheets can be sealed with adhesive tape, but this is not necessary if the sheets are bedded to the frames with some sealing cement, which should be applied generously. The dome, shown in the photograph (Plate 14, top), has now been in action for some years, and works excellently.

A design somewhat similar to the foregoing has been adopted by Commander H. R. Hatfield for his observatory at Sevenoaks in Kent. On account of its appearance he has christened it the 'Beehive', and he describes it thus:

'The Beehive is built on a slope of about one in five, and is supported on six brick pillars. The main floor bearers are 6 × 2in, and the minor ones are 4 × 2in.

'The bottom half of the Beehive is hexagonal with 6ft sides, made of tongued and grooved cedar on a soft wood framing. It supports the shelf made of ⅜in marine plywood, on which the top rotates. The shelf is covered with thin aluminium sheet to take the wear of the Shepheard castors which run on it. The top half, which rotates, is twelve-sided, and made of tongued-and-grooved cedar on soft wood framing. There are two windows, and a hinged door comprises one of the twelve sides.

'A 3ft wide roof section slides up and back on a track. Originally it just slid on greased runners, but now it has fitted rollers; both are satisfactory. The roof is of shingles; they take a long time to fit, but are well worth it.

'The Beehive is now some four-and-a-half-years old, and is still completely dry and has given no trouble. Cedar is expensive,

Q

but worth it. It is very nice to work with, and is easy to preserve. Shingle roofs last longer than human beings! The total cost of materials was about £60, and it was built during the summer of 1965 by one man working evenings and week-ends.

'The top rotates on twelve Shepheard castors, one for each section. To stop it falling off, six horizontal wheels about 4in in diameter, fitted to the six-sided bottom half, bear against a circular wooden rail which is fitted inside the bottom of the top half. Power and light are laid on, and the instrument housed is a 12in reflector on a German mounting. The rotating part weighs about 5cwt. Although it is lashed down in strong winds, it has never shown any tendency to take off. The sliding section of the roof is held down on the track when it is open (up) by the tackle which is used to pull it open.' (Fig 60.)

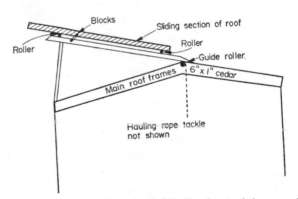

Fig 60   *Commander Hatfield's 'beehive' sliding roof*

While not strictly a dome, this design is obviously one of considerable note, and may be regarded as an efficient and ingenious compromise. An overall view of the 'Beehive' is shown in Plate 15.

Another variant of this design is that of Patrick Moore's observatory (see Plates 12 and 13), already mentioned in the Introduction. The method of opening the slit is quite different from that of the 'Beehive', but has proved very effective. The lower part of the revolving portion is glazed. Remember, however, that without adequate ventilation the glass will cause a greenhouse effect, and the interior temperature will rise during the day when the weather is sunny, thus heating the instrument and causing bad visibility if observation is begun with the instrument and observatory interior at a higher temperature than the outside air. No trouble of this sort seems to have been experienced, but the point must be kept in mind when designing an observatory of this type.

It will be well appreciated that there are many variants of design that can be employed, such as the number of sides to the observatory and the whole method of construction, as well as the method of providing a satisfactory cover to the opening section. A cheaper form of rotating drum can be designed instead of the dome type of observatory, as instanced in an example designed by Alan Wake of Teignmouth in Devon.

In this case the frame is made of 2 × 2in timber covered with hardboard. The design is as simple as possible, and comprises nine sides, each 4ft 6in wide. The overall height is only 6ft 6in. The roof members are only 2 × 1in timber, but experience indicates that they should be stronger: say 3 × 1in. The overall diameter is 12ft, with the inner floor diameter 10ft and the roof, raised at the centre, reaching a maximum height of 9ft 10in at the apex. The opening hatches are 4ft 6in wide at the wall, decreasing to 2ft at the apex. The hatch covers, two in number, are hinged and held rigid in the open position by two metal arms which can be placed in suitable fastenings after the hatches are opened.

The concrete wheel track was cast circular to accommodate the wheels of the sack truck type, rubber-tyred and with roller bearings. The whole building thus rotates in a manner similar to that of the author's original observatory. In Wake's case the concrete track stands 12in high above the ground. The track to accommodate the wheels was cut by using a boom pivoted at the centre of the circular track, and the track itself was cut before the concrete set hard. The wheels are held by blocks with 2in spacers between them, so that the wheels can rotate on their axles bolted underneath the blocks. The drum accommodates a 15in (40cm) reflector working at f/3·9.

Yet another efficient and attractive design is that of A. C. Curtis in Winchester, illustrated in the photograph, Plate 16, bottom. The dome is sixteen-sided, made of hardwood to avoid shrinkage or warping. The top and bottom rails are of teak, and the boarding is of mahogany. The framework is bolted together, and has produced an extremely stable building without the necessity of bracing.

The rail on which the dome rotates is of $1\frac{1}{2} \times 1\frac{1}{2} \times 3/16$ angle-iron, prepared at the rolling mill in six segments to form the complete circle. One 3in V pulley wheel is fixed to each of the sixteen sections by two $\frac{1}{4}$in bolts. The frames are covered by hardboard, with a further layer of Ruberoid stuck on the outside for weather protection.

Perhaps the outstanding feature of this observatory is the arrangement of the opening shutter. This is made by an adaptation of the design of one of the square-cut frames, resting in rebates formed by screwing slats to the sides of the adjoining frames.

Being covered with aluminium sheeting, this shutter is light enough to be lifted off for observing sessions. Clear opening of the shutter is 2ft 6in, and extends to about 10° from the zenith. This zenithal area is covered by a lid which can be removed on the infrequent occasions which it is found necessary. Normally, the zenith can be reached by rotating the dome to a suitable position.

This scheme of making the shutter entirely removable does overcome the difficulties of ensuring that the shutter is weathertight. It also simplifies the construction. However, for obvious reasons, it is strictly limited to domes of smaller sizes. Mr Curtis dome is 11ft in outside diameter and stands 9ft high.

It is essential that the floor of the observatory should not touch the instrument pedestal. There is always some vibration in a wooden floor, and people walking thereon can easily destroy the performance of a telescope by shaking it. If the floor is of concrete of reasonably deep construction, then it may be possible to cement the telescope pier into the floor as one structure without encountering vibration troubles.

A wooden floor is warmer to the feet than a concrete one, and is preferred by the author. It is perhaps unnecessary to emphasise the point, but it must be remembered that a wood floor needs ventilation under it to prevent rot. A method often employed is to have a dry, well-drained and ventilated space between the bottom of the wooden floor and the ground. This may be surfaced with dry cinders, etc, to avoid collection of moisture, but should in any case be provided with plenty of ventilation. As is shown in Plate 16, bottom, the bricks on which the dome rests are well spaced, with plenty of openings between for underfloor ventilation.

In the case of concrete-floored or wholly rotatable buildings, this ventilation, necessary to avoid condensation following the rise of outside temperature, must be arranged differently. The wholly revolving type is separate from the actual floor, and the gap may be sealed by fixing a flap or curtain of canvas on the outside of the building in such a way that air is allowed to enter slowly, but if a strong wind hits the curtain it is closed like a flap valve against any sudden gust. This method is very efficient in keeping the temperature inside the observatory near to that outside, and adds to observing comfort. The concrete-floored type equally needs proper ventilation. With Patrick Moore's dome there is a slight gap between the top part of the wooden structure and the ledge at the bottom of the glazed portion which rotates. This allows air circulation, and may be one reason why no trouble has been found with turbulence.

As alternatives to the materials usually employed, and described in this chapter, it is of course quite possible to make observatory domes in either metal or fibreglass, or similar plastic material. The techniques are in each case particular to the actual material used, and each has its attractions and its disadvantages. With fibreglass or plastic construction, it may be possible to do without the framing of the dome, and to use the moulded sections alone.

Metal work is normally erected on a frame, which is set up first, and the cladding metal sheets are put on afterwards. Care must be taken in this design to allow adequate and continuous ventilation of the dome, since metal heats up readily in sunshine and the heat is conveyed to the inside of the observatory. Perhaps a double skin may be employed, or the inside of the dome may be drawn with some heat-resisting material similar to asbestos sheeting. Much depends on the amount of money available, the labour required, and the experience or otherwise of the constructors in the use of the various materials.

Most people are acquainted with woodwork in some way or another. Therefore, emphasis has been put on this type of dome in this chapter. However, provided that one can handle the other materials and is acquainted with the techniques involved, there is no reason why they should not be used. Neither would there be any objection to using more than one type of material, say a metal frame with hardboard cladding.

From the examples described and illustrated in this chapter, it will be appreciated that the possible variations in design are virtually infinite. Each constructor can supply some detail or variation applicable to his own needs and wishes. It should also be noted that designs can be relatively simple and cheap, or, alternatively, more complicated and costly. It is up to the designer to make his choice.

With these two principles in mind, it will be appreciated that there is really no need for observers who wish to have more comfort, and thereby greater accuracy in their work, to deny themselves a dome or something similar.

The building need not take up much space in the garden, and it need not be unsightly. A nicely-designed observatory, properly landscaped into its surroundings, can even look attractive.

Contributors

IAIN NICOLSON is lecturer in astronomy at the Hatfield Polytechnic. He is a graduate of St Andrews University, and is associate editor of the quarterly periodical *Astronomy & Space*.

G. A. HOLE has a world reputation as a maker of astronomical optics, and is the founder of the optical form of George Hole & Son (Brighton, Sussex). He is director of the Instruments and Observing Methods Section of the British Astronomical Association.

T. W. RACKHAM was formerly at Cambridge; between 1968 and 1971 he was director of the Armagh Planetarium in Northern Ireland, and is now at Jodrell Bank.

TERENCE MOSELEY is an amateur astronomer with his own observatory in Armagh. He graduated in psychology from the Queen's University, Belfast, and is associate editor of *Astronomy & Space*.

H. K. ROBIN, CBE, has his observatory at Tunbridge Wells, in Kent, and is a council member of the British Astronomical Association.

GILBERT E. SATTERTHWAITE was formerly a professional astronomer at the Royal Greenwich Observatory, and is now engaged in scientific publishing with Pitman's. He is director of the Saturn section of the British Astronomical Association.

H. E. DALL is well known in the field of scientific optics. He has been responsible for fundamental advances (such as the Dall-Kirkham telescope). He lives in Luton.

J. C. D. MARSH is senior lecturer in astronomy at the Hatfield Polytechnic. He is particularly interested in infra-red astronomy and is at present concerned with the establishment of a major research station in Teneriffe.

COLIN A. RONAN was formerly on the secretariat of the Royal Society, and is concerned mainly with the history of science and with scientific photography. He is the author of many books, and is editor of the Journal of the British Astronomical Association.

COMMANDER H. R. HATFIELD, R.N., is a serving officer in the Royal Navy. He has established a world reputation as an astronomical photographer, and is the author of a well-known lunar photographic atlas. He is vice-president of the British Astronomical Association.

F. R. SPRY lives at Selsey, in Sussex, where he has his private observatory. His interests lie largely in telescope and observatory construction.

J. HEDLEY ROBINSON retired from a career in banking some years ago, and now devotes much of his time to astronomy; he has his observatory at Teignmouth, in Devon. He is director of the Mercury and Venus section of the British Astronomical Association.

PATRICK MOORE, OBE, is an amateur astronomer living at Selsey. He is director of the Lunar section of the British Astronomical Association, and editor of the periodical *Astronomy & Space*. He was director of the Armagh Planetarium from 1965 to 1968.

DR L. WILSON is carrying out research at the University of Lancaster He is a specialist in studies of the Moon as well as in all aspects of astronomical photography.

References

Chapter 1   Moore, Patrick. *The Amateur Astronomer* 1971
Norton, A. P. *Star Atlas* 1972

Chapter 2   Roth, G. D. *The Amateur Astronomer and his Telescope* 1972
Ingalls, Albert G. (Ed). *Amateur Telescope Making*, Vols 1 and 2

Chapter 5   Sidgwick, J. B. *Amateur Astronomer's Handbook* 1971

Chapter 6   Sidgwick, J. B. *Amateur Astronomer's Handbook*, pp 446 et seq 1971

Chapter 8   Longhurst, R. S. *Geometric and Physical Optics*

Chapter 9   Clarke, D. 'Studies in Astronomical Polarimetry', *MNRAS*, 130, Vol. 82, 1965
Hiltner, W. A. (Ed). *Stars and Stellar Systems, Vol. 11, Astronomical Techniques*, pp 178 et seq 1966
Hiltner, W. A. (Ed). *Stars and Stellar Systems, Vol. 11, Astronomical Techniques*, pp 229 et seq 1966
Sartory, P. K. 'A Method of Rendering Obvious Small Differences in Colour or Contrast Observations', *JBAA*, 75, no 2, 98, 1965
Sidgwick, J. B. *Amateur Astronomer's Handbook*, p 377 1971
Sidgwick, J. B. *Amateur Astronomer's Handbook*, p 297 et seq 1971

Chapter 12   Ball, R. S. *A Popular Guide to the Heavens*, pp 54–5 1957 (1)
Dall, H. E. *Practical Amateur Astronomy*, ch 5 (Ed P. Moore) 1969 (2)

Fielder, G. *MNRAS, 118*, 547–50, 1958 (3)

MacDonald, T. L. *JBAA, 41*, 367–79, 1931 (4)

McMath, R. R., Petrie, R. M., Sawyer, H. E. *Publ Obs Univ Mich, 6*, 67–76, 1937 (5)

Rackham, T. W. *Astronomical Photography at the Telescope* (6)

Radio Society of Great Britain, *The Amateur Radio Handbook* (7)

Index